Intelligent Control and Learning Systems

Volume 11

Series Editor

Dong Shen, School of Mathematics, Renmin University of China, Beijing, China

The Springer book series Intelligent Control and Learning Systems addresses the emerging advances in intelligent control and learning systems from both mathematical theory and engineering application perspectives. It is a series of monographs and contributed volumes focusing on the in-depth exploration of learning theory in control such as iterative learning, machine learning, deep learning, and others sharing the learning concept, and their corresponding intelligent system frameworks in engineering applications. This series is featured by the comprehensive understanding and practical application of learning mechanisms. This book series involves applications in industrial engineering, control engineering, and material engineering, etc.

The Intelligent Control and Learning System book series promotes the exchange of emerging theory and technology of intelligent control and learning systems between academia and industry. It aims to provide a timely reflection of the advances in intelligent control and learning systems. This book series is distinguished by the combination of the system theory and emerging topics such as machine learning, artificial intelligence, and big data. As a collection, this book series provides valuable resources to a wide audience in academia, the engineering research community, industry and anyone else looking to expand their knowledge in intelligent control and learning systems.

Qikun Shen

Distributed Fault-Tolerant Consensus Control of Leader-Following Systems

Based on Adaptive Control Approach

Springer

Qikun Shen
College of Information Engineering
Yangzhou University
Yangzhou, China

ISSN 2662-5458 ISSN 2662-5466 (electronic)
Intelligent Control and Learning Systems
ISBN 978-981-99-7425-2 ISBN 978-981-99-7426-9 (eBook)
https://doi.org/10.1007/978-981-99-7426-9

This Springer imprint is published by the registered company Springer Nature Singapore Pte Ltd.
The registered company address is: 152 Beach Road, #21-01/04 Gateway East, Singapore 189721, Singapore

To our families for their love and support

Preface

For multi-agent systems, their distributed consensus control problems have attracted widespread attention in the past decades, and the attention is still increasing now. The main reason for the attention is that the multi-agent systems have been widely applied in various fields, such as power systems, unmanned vehicles, mobile robots, and so on. Among the researches of multi-agent systems, consensus is common but fundamental although it has been deeply investigated. In general, it is divided into two types. One is leaderless consensus, and the other is named leader-following consensus. Up to now, for the consensus control, abundant efforts have been made by experts and scholars in various fields, and lots of results have been obtained.

On the other hand, with the development of science and technology, the multi-agent systems considered in this book and the other modern control systems become increasingly complex, and the number of actuators, sensors, and other system components involved in the systems is increasing. However, in the practical application, it is inevitable that the components may become deteriorating or unhealthy, which means that some faults occurred in the components and they become faulty. The faults prevent the system components from working in a normal manner. What's worst, the faulty components cannot work at all. Obviously, system performance will be deteriorated, and instability even or catastrophic accidents may occurr. Hence, to improve or enhance system reliability, fault-tolerant control (FTC) techniques and approaches have been proposed by experts and scholars from various fields. For single-agent systems, among the system components involved in the controlled systems, actuators, and sensors are more important and also often become faulty. However, for multi-agent systems (MASs) including leader following considered in this book, the situation becomes more complicated. For the MASs, the faults occur in not only actuators and sensors but also the components in communication networks.

In recent years, many researchers pay attention to the multi-agent systems with faults, and various distributed fault-tolerant control methods have been proposed for dynamic systems including linear and nonlinear systems, certain and uncertain systems. For the faults that occurred in multi-agent systems, however, most of the existing results reported in literature focus on actuator and/or sensor faults. In fact, there exist other faults such as communication faults in multi-agent systems. Up to now, for the fault-tolerant control design for multi-agent systems including leader-following systems, although fruitful results have been obtained, the theoretical studies are not perfect, and there also are many problems, which are worth to be further deeply investigated due to its academic meaning as well as its practical one.

Focused on the distributed adaptive consensus control problems for leader-following systems with faults, the book is organized into the following chapters.

Chapter 1 introduces the technical problems faced by distributed control of multi-agent systems, especially distributed fault-tolerant control. Firstly, fault types in multi-agent systems are introduced. Based on the location of the fault and the impact on system performance, the faults are divided into two types: local and global faults. Then fault-tolerant control methods and results against the faults are introduced. Finally, the organization of the book is given.

In Chap. 2, cooperative tracking control is investigated for a class of leader-following systems under a weighted directed graph with a fixed topology, and a novel distributed control method is constructed such that each follower in the system synchronizes asymptotically to a leader and the tracking error converges to a small adjustable neighborhood of the origin in spite of the presentence of actuator faults. By using Lyapunov stability theory and algebraic graph theory, stability analysis and parameter convergence of the resulting closed-loop system are conducted. Comparing with the existing results in literature, the control method in this work not only minimizes the time delay between fault accommodation and occurrence but also reduces its adverse effect of the faults impacted on control performance. Furthermore, the fault-tolerant control does not require fault isolation model, which is necessary in the traditional active fault-tolerant scheme.

In Chap. 3, an output consensus control design is considered for a class of nonlinear leader-following systems, which is modeled as a directed graph. It is assumed that the topology of the directed graph is fixed and each follower in the graph is subject to dead-zone and saturation. Combining backstepping control technology with the fuzzy-logic systems' function approximation capability, an adaptive fuzzy distributed control method is designed to ensure that each follower in the graph synchronizes asymptotically to the leader and the synchronization errors converge to the origin in spite of external disturbances and unknown system dynamics. Based on Lyapunov stability theory and basic graph theory, stability analysis and proof of the resulting closed-loop system are developed.

In Chap. 4, a consensus problem is considered for multi-agent systems, which contain one leader and multiple followers and is modeled a directed graph with a fixed topology. In the systems, the actuators of followers are assumed to be subject to unknown backlash-like hysteresis. Based on Lyapunov stability theory and basic graph theory, combining variable structure control technology with the neural networks' function approximation capability, a novel distributed adaptive control scheme is proposed to guarantee that each follower in the graph synchronizes asymptotically to the leader with the synchronizing errors being bounded. Note that, for neural networks, they can approximate as accurately as possible any continuous function just under the condition that the variables in the function should be within a compact sets. In the scheme, the supervisor control strategy first guarantees the necessary condition, namely, the signals in the resulting closed-loop system are ensured to always be within the corresponding compact sets, respectively. Furthermore, the effects of approximation and modeling errors are reduced by adopting suitable compensation terms.

In Chap. 5, the output consensus is discussed for a class of second-order uncertain nonlinear leader-following systems described by a directed graph, where the actuator of each follower is subject to an unknown nonlinear dead-zone and the topology of the graph is fixed. By the function approximation capability of fuzzy logic systems, a distributed adaptive fuzzy control method is proposed to guarantee that output synchronization between the leader and followers is achieved. From Lyapunov stability and graph theory, it is proven that all the signals of the closed-loop system including synchronizing errors are semi-globally uniformly and ultimately bounded.

In Chap. 6, the synchronization problem is considered for a class of nonlinear leader-following systems with transmission faults. Firstly, the transmission faults are modeled. Then, using adaptive control and the function approximation capability of neural networks, two distributed adaptive control schemes are proposed, respectively, for the two faults to guarantee that each follower synchronizes asymptotically to a leader and the synchronizing errors are cooperatively uniformly and ultimately bounded. In this scheme, the two transmission faults are compensated, respectively. Based on algebraic graph theory and Lyapunov stability theory, the closed-loop system stability analysis is conducted.

Chapter 7 considers the synchronization problem of a class of nonlinear leader-following systems with signal transmission faults. By using adaptive control and the function approximation capability of neural networks, an adaptive fault-tolerant control method is constructed to obtain synchronization between the leader and followers. In this work, the signal transmission faults have been compensated by using suitable fault compensation algorithms. Based on Lyapunov theory and basic graph theory, it is proven that each follower asymptotically synchronizes to the leader with the synchronizing errors converging to a small neighborhood of the origin despite the signal transmission faults occurring in sensors.

In Chap. 8, the consensus control problem is investigated for a class of nonlinear leader-following systems with sensor faults, and a distributed adaptive neural fault-tolerant backstepping control method is proposed to guarantee the output consensus between the leader and followers. Note that, because the faults occur in the input of neural network used to approximate unknown continuous functions, the function approximation abilities of neural networks have decreased. In this method, the problem of declining function approximation ability has been solved by compensating for the sensor faults. Furthermore, an observer/filter is constructed to estimate the unmeasurable states. From Lyapunov stability theory, it is proved that the closed-loop system is uniformly ultimate boundness with the tracking errors converging to a small adjustable neighborhood of the origin.

In Chap. 9, several future research directions are predicated.

This book provides recent theoretical results and applications of the consensus control for nonlinear leader-following systems. Combining adaptive control techniques, fuzzy logic systems, and neural networks with other control techniques or approaches, this book investigates the consensus control problem of leader-following systems and proposes the corresponding control scheme. This book intends to provide the readers with a good understanding on consensus control based on adaptive control technology. This book can serve as a reference for the main research issues and results on nonlinear multi-agent systems including leader-following systems for researchers devoting to various areas of control theory, as well as a material for graduate and undergraduate students interested in nonlinear multi-agent systems including leader-following systems and their applications. Some prerequisites for reading this book include nonlinear system theory, matrix theory, mathematics, basic graph theory, and so on.

Yangzhou, China
July 2023

Qikun Shen

Acknowledgements The contents included in this book are an outgrowth of our academic research activities over the past several years. This book was partially supported by the Publication Foundation of Yangzhou University, partially supported by the National Natural Science Foundation of China (62273300), and also supported by the Publication Foundation of College of Information Engineering, Yangzhou University.

Contents

Acronyms

$\mathfrak{R}$	The field of real numbers
$\mathfrak{R}^r$	The r-dimensional real vector space
$\lvert\cdot\rvert$	The Euclidean norm
$\mathscr{L}_1$	$a(t) \in \mathscr{L}_1$ if $\int_0^\infty \lvert a(t)\rvert dt < \infty$
$\mathscr{C}^k$	The set of k times continuously differentiable functions
$\lVert\cdot\rVert_{[a,b]}$	The supremum norm of a signal on the time interval $[a, b]$
Class $\mathscr{K}$	A class of strictly increasing and continuous functions $[0, \infty) \to [0, \infty)$ which are zero at zero
Class $\mathscr{K}_\infty$	The subset of $\mathscr{K}$ consisting of all those functions that are unbounded
Class $\mathscr{K}\mathscr{L}$	$\beta : [0, \infty) \times [0, \infty) \to [0, \infty)$ belongs to class $\mathscr{K}\mathscr{L}$ if $\beta(\cdot, t)$ is of class $\mathscr{K}$ for each fixed $t \geq 0$ and $\beta(s, t) \to 0$ as $t \to \infty$ for each fixed $s \geq 0$
$\lambda_{\max}(\cdot)$	The maximal eigenvalue
$\lambda_{\min}(\cdot)$	The minimal eigenvalue
$L_g h$	The Lie derivative of h along a vector field g
t^-	The left limit time instant of t
$(\cdot)^\top$	The transposition
$\forall$	For all
$\exists$	There exists
$\subset$	Subset of
$\subset$	Belongs to
$\triangleq$	Define
$\cap$	Intersect
$\cup$	Union
$\sum$	Sum
$\prod$	Product
FTC	Fault-tolerant control

Chapter 1
Introduction

For multi-agent systems, their distributed consensus control problems have attracted widespread attention in the past decades, and the attention is still increasing now. The main reason for the attention is that the multi-agent systems have been widely applied in various fields, such as power systems, unmanned vehicles, mobile robots, and son Defoort et al. (2008); Cruz et al. (2007); Wang et al. (2007); Chen et al. (2010); Murray (2007); Ren and Cao (2011); Olfati-Saber et al. (2007); Ren et al. (2007); Hung et al. (2012); Mathiyalagan et al. (2014); Zhao et al. (2014); Nguyen and Tran (2013); Szupiluk et al. (2014); Sedziwy (2014); Shen and Shi (2015); Du et al. (2012); Chen et al. (2014); Yu and Xia (2012); Ren et al. (2007); Zhang et al. (2011); Zhang and Lewis (2012); Zhang et al. (2012). Among the researches of multi-agent systems, consensus is common but fundamental although it has been deeply investigated. In general, it is divided into two types. One is leaderless consensus, and the other is named as leader-following consensus. Up to now, for the consensus control, abundant efforts have made by experts and scholars in various fields, and lots of results have been obtained Shen et al. (2014); Li et al. (2014); Zhu and Cheng (2010); Wang and Slotine (2006); Hong et al. (2008); Ni and Cheng (2010); Guo et al. (2013).

On the other hand, with the development of science and technology, the multi-agent systems considered in this book and the other modern control systems become increasingly complex, and the number of actuators, sensors, and the other system components involved in the systems is increasing Chen and Patton (1999); Mahmoud et al. (2003); Yang et al. (2010); Wang et al. (2013); Du et al. (2015). However, in the practical application, it is inevitable that the components may become deteriorating or unhealthy, which means that some faults occurred in the components and they become faulty. The faults prevent the system components from working in a normal manner. What's worst, the faulty components cannot work at all. Obviously, system performance will be deteriorated, and even instability or catastrophic accidents may occur. Hence, to improve or enhance system reliability, fault-tolerant control (FTC) techniques and approaches including fault diagnosis (FD) Shen et al. (2012a); Gertler (1988) have been proposed by experts and scholars from various fields. For

Q. Shen, *Distributed Fault-Tolerant Consensus Control of Leader-Following Systems*, Intelligent Control and Learning Systems 11,
https://doi.org/10.1007/978-981-99-7426-9_1

the traditional single-agent systems, among the system components involved in the controlled systems, actuators and sensors are more important and also often become faulty Shen et al. (2012a, b, 2013a, b); Gong et al. (2019); Gertler (1988). However for multi-agent systems (MASs) including leader-following considered in this book, the situation becomes more complicated Shen et al. (2014, 2021, 2017); Shen and Shi (2016). For the MASs, the faults occur in not only actuators and sensors Shen et al. (2014) but also the components in communication networks Fu et al. (2020); Gong et al. (2020); Shen et al. (2019); Gong et al. (2019).

In recent years, many researchers pay attention to the multi-agent systems with faults including input nonlinearities regarded as special actuator faults in this book, and various distributed fault-tolerant control methods have been proposed for multi-agent systems including linear and nonlinear MASs, certain and uncertain MASs Shen et al. (2014, 2021); Fu et al. (2020); Gong et al. (2020); Shen et al. (2019); Gong et al. (2019); Shen et al. (2017); Shen and Shi (2016). For the faults occurred in multi-agent systems, however, most of the existing results reported in literature focus on actuator and/or sensor faults. In fact, there exist the other faults such as communication faults in multi-agent systems. Up to now, for the fault-tolerant control design for multi-agent systems including leader-following systems, although fruitful results have been obtained, the theoretical research is not perfect, and there also exist many issues, which are worth to be further studied deeply because of their academic value as well as its practical one.

1.1 Fault Types in MASs: Local and Global Faults

It is well known that, for single-agent systems, faults occur only inside the system. The faults are named as actuator faults, sensor faults, and so on Frank and Seliger (1990); Frank (1996); Jafari et al. (2022); Zhang et al. (2020); Gong et al. (2019); Frank (1994); Zheng and Shen (2023); Song et al. (2002); Shen et al. (2014); Vidyasagar and Viswanadham (1985); Xu and Shen (2022); Saeks and Murray (1982); Kabamba and Yang (1991); Jiao and Shen (2020); Morari (1985); Chen et al. (2019); Shen et al. (2008); Yu and Fan (1990); Bao et al. (2003); Zhang et al. (2002); Saljak (1980); Kaminer et al. (1995); Deng et al. (2020); Moerder (1989); Ma et al. (2023); Srichande and Walker (1993); Ranmamurthi and Agogino (1993); Wu et al. (2000); MorseW and Ossman (1990); Huang and Stengel (1990); Napolitanob and Swaim (1991); Kwong et al. (1995); Tao et al. (2002); Jin and Yang (2009); Jiang et al. (2006); He et al. (2009); Liu et al. (2011). Different from the single-agent systems, the faults in multi-agent systems may occur not only inside single-agent systems but also outside single-agent systems. That is to say, not only the actuators, sensor, and the other system components in single-agent systems may become faulty, but also other parts outside single-agent systems, for example, communication links in communication network, maybe faulty in the practical applications. Obviously, the faults occurred in multi-agents including leader-following systems are more diversified, which will bring more challenges to the fault-tolerant control designing of multi-agent systems.

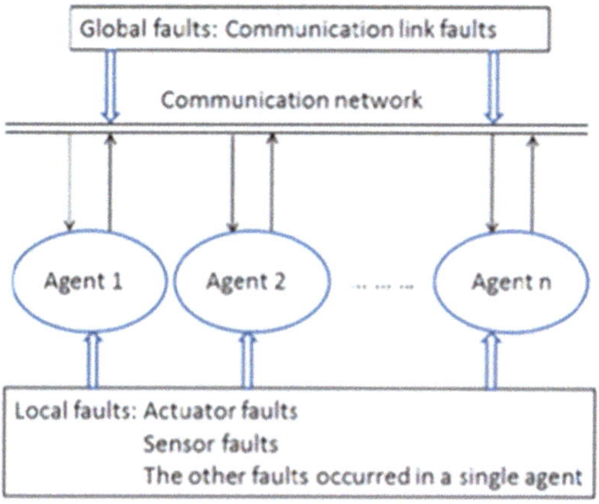

Fig. 1.1 Fault types occurred in multi-agent systems

For a fault in multi-agent systems, according to the location of the fault and/or its impact on the controlled system, the fault can be divided into two categories: local fault and global fault, which is shown in Fig. 1.1.

- **Local faults** are the faults occurring only inside single systems (agents). The classic actuator and sensor faults belong to local faults. In addition, the faults in the other system components also belong to the local faults.
- **Global faults** are the faults occurring outside single agents. It is via communication network that all agents in the whole multi-agent systems exchange their information. Note that the network does not belong to one of the agents and are outside agents. Hence, the faults occurred outsider agents, for example, the faults in communication links belong to the global faults.

In order to remove or reduce the influence of the faults on the controlled systems, it is necessary to design the corresponding fault diagnosis, compensation algorithms, and fault-tolerant control schemes Dong and Yang (2008); Wang et al. (2008); Li and Tao (2009); Du et al. (2011); Wu (2004); Wang and Wen (2010); Zhang and Guay (2003); Zhang et al. (2004, 2002); Wang and Wen (2011); Wu et al. (2010); Richard (2003); Hu et al. (2021); You et al. (2006); Gao and Zhang (2006); Fridman and Shaked (2002); Liu and Zhou (2003); Zhao et al. (2010); Zhang et al. (2009); Bai et al (2007); Shin et al. (2004); Belcastro (2006); Hu et al. (2011); Hu and Xiao (2011). Although there are abundant results on fault-tolerant control for multi-agent systems, most of the results focus on the local faults, and the global faults are not investigated deeply. From the above-mentioned definitions of local and global faults, we know it is obvious that the local and global faults have different impacts on the performance of multi-agents. What's more important, the compensation algorithm designing of the local and global faults is different. Therefore, for multi-agent systems, the first task is how to distinguish the faults and decide their types. Then, the corresponding fault compensation algorithms and accommodation schemes should be designed to compensate for the faults, respectively.

1.2 Fault-Tolerant Control Against Local Faults

For multi-agent systems including leader-following systems, common local faults are actuator and sensor faults. To reduce or relax their influence impacted on the controlled systems, researchers widely investigate the faults and construct various fault-tolerant control methods for different dynamic systems including linear and nonlinear systems, certain and uncertain systems, the systems with or without time delay, and so on Shen et al. (2014, 2021, 2017); Shen and Shi (2016).

On the other hand, input nonlinearity is also common in the practical applications. In fact, saturation, dead-zone, sector nonlinearity, and other input nonlinearities, as inherent characteristics of the actuators, are often regarded as special actuator faults. By the way, the authors also see the nonlinearities as the actuator faults in this book. In past 10 years, they have been studied by many researchers, and by using many techniques available, various fault-tolerant control methods have been designed to remove or reduce the effects of the nonlinearities Defoort et al. (2008); Cruz et al. (2007); Wang et al. (2007); Chen et al. (2010); Murray (2007); Ren and Cao (2011); Olfati-Saber et al. (2007); Ren et al. (2007); Hung et al. (2012); Chen and Patton (1999); Mahmoud et al. (2003); Yang et al. (2010); Wang et al. (2013).

Note that the above-mentioned results about fault-tolerant control against the faults including the input nonlinearities in Defoort et al. (2008); Cruz et al. (2007); Wang et al. (2007); Chen et al. (2010); Murray (2007); Ren and Cao (2011); Olfati-Saber et al. (2007); Ren et al. (2007); Hung et al. (2012); Chen and Patton (1999); Mahmoud et al. (2003); Yang et al. (2010); Wang et al. (2013) focused on only single-agent systems and MASs were not considered. Hence, the FTC problem of MASs is still open and challenging, which is worthy to be widely studied.

1.3 Fault-Tolerant Control Against Global Faults

For the consensus control problem of leader-following systems, a leader actually acts as a command generator to generate the desired reference trajectory and ignore the information of the follower agents in the transmission network. Meanwhile, each follower agent only utilizes the information of itself and its neighbors to design the so-called distributed controller, which guarantees the consensus between itself and the leader. That is to say, the follower agent can have access to its neighbors' information via transmission network. However, in the works about consensus control, the transmission nonlinearity or fault is seldom considered Sedziwy (2014); Shen and Shi (2015); Shen et al. (2014); Du et al. (2012); Chen et al. (2014); Li et al. (2014). In fact, the transmitted information among agents are always subject to various constrains, such as sensor noise or fault, channel fading, quantization errors, and so on Fu et al. (2020); Gong et al. (2020); Shen et al. (2019); Gong et al. (2019). These constrains cause the transmission nonlinearities. Hence, how to design an proper distributed control scheme for the leader-following systems with the transmission nonlinearities naturally becomes more important and challenging.

1.4 Organization of the Book

This book presents several fundamental problems of fault-tolerant consensus control for leader-following systems. Combining adaptive control technique with the other control techniques or approaches, a basic theoretical framework is formed towards the issues of consensus control of leader-following systems. In addition, to conveniently read this book, some preliminaries including same or similar lemmas are introduced in different chapters. This book contains nine chapters, which exploit several independent yet related topics in the detail.

In Chap. 2, cooperative tracking control is investigated for a class of leader-following systems under a weighted directed graph with a fixed topology, and a novel distributed control method is constructed such that each follower in the system synchronizes asymptotically to a leader and the tracking error converges to a small adjustable neighborhood of the origin in spite of the presence of actuator faults. By using Lyapunov stability theory and algebraic graph theory, stability analysis and parameter convergence of the resulting closed-loop system are conducted. Comparing with the existing results in literature, the control method in this work not only minimizes the time delay between fault accommodation and occurrence but also reduces its adverse effect of the faults impacted on control performance. Furthermore, the fault-tolerant control does not require fault isolation model, which is necessary in the traditional active fault-tolerant scheme.

In Chap. 3, an output consensus control design is considered for a class of nonlinear leader-following systems, which is modeled as a directed graph. It is assumed that the topology of the directed graph is fixed and each follower in the graph is subject to dead-zone and saturation. Combining backstepping control technology with the fuzzy-logic-systems' function approximation capability, an adaptive fuzzy distributed control method is designed to ensure that each follower in the graph synchronizes asymptotically to the leader and the synchronization errors converge to the origin in spite of external disturbances and unknown system dynamics. Based on Lyapunov stability theory and basic graph theory, stability analysis and proof of the resulting closed-loop system are developed.

In Chap. 4, a consensus problem is considered for multi-agent systems, which contain one leader and multiple followers and is modeled as an directed graph with a fixed topology. In the systems, the actuators of followers are assumed to be subject to unknown backlash-like hysteresis. Based on Lyapunov stability theory and basic graph theory, combining variable structure control technology with the neural-networks' function approximation capability, a novel distributed adaptive control scheme is proposed to guarantee that each follower in the graph synchronizes asymptotically to the leader with the synchronizing errors being bounded. Note that, for neural networks, they can approximate as accurately as possible any continuous function just under the condition that the variables in the function should be within a compact set. In the scheme, the supervisor control strategy firstly guarantees the necessary condition, namely, the signals in the resulting closed-loop system are ensured to always be within the corresponding compact sets, respectively. Furthermore, the effects of approximation and modeling errors are reduced by adopting suitable compensation terms.

In Chap. 5, the output consensus is discussed for a class of second-order uncertain nonlinear leader-following systems described by a directed graph, where the actuator of each follower is subject to unknown nonlinear dead-zone and the topology of the graph is fixed. By the function approximation capability of fuzzy logic systems, a distributed adaptive fuzzy control method is proposed to guarantee that output synchronization between the leader and followers is achieved. From Lyapunov stability and graph theory, it is proven that all the signals of the closed-loop system including synchronizing errors are semi-globally uniformly ultimately bounded.

In Chap. 6, the synchronization problem is considered for a class of nonlinear leader-following systems with transmission faults. Firstly, the transmission faults are modeled. Then, using adaptive control and the function approximation capability of neural networks, two distributed adaptive control schemes are proposed, respectively, for the two faults to guarantee that each follower synchronizes asymptotically to a leader and the synchronizing errors are cooperatively uniformly and ultimately bounded. In this scheme, the two transmission faults are compensated, respectively. Based on algebraic graph theory and Lyapunov stability theory, the closed-loop system stability analysis is conducted.

Chapter 7 considers the synchronization problem of a class of nonlinear leader-following systems with signal transmission faults. By using adaptive control and the function approximation capability of neural networks, an adaptive fault-tolerant control method is constructed to obtain the synchronization between the leader and followers. In this work, the signal transmission faults have been compensated by using suitable fault compensation algorithms. Based on Lyapunov theory and basic graph theory, it is proven that each follower asymptotically synchronizes to the leader with the synchronizing errors converging to a small neighborhood of the origin despite the signal transmission faults occur in sensors.

In Chap. 8, the consensus control problem is investigated for a class of nonlinear leader-following systems with sensor faults, and a distributed adaptive neural fault-tolerant backstepping control method is proposed to guarantee the output consensus between the leader and followers. Note that, because the faults occur in the input of neural network used to approximate unknown continuous functions, the function approximation abilities of neural networks have decreased. In this method, the problem of declining function approximation ability has been solved by compensating for the sensor faults. Furthermore, an observer/filter is constructed to estimate the unmeasurable states. From Lyapunov stability theory, it is proved that the closed-loop system is uniformly ultimate boundness with the tracking errors converging to a small adjustable neighborhood of the origin.

In Chap. 9, several future research directions are predicated.

References

Bai L, Tian Z, Shi S (2007) Robust fault detection for a class of nonlinear time-delay systems. J Frankl Inst 344(6):873–888

Bao J, Zhang WZ, Lee PL (2003) Decentralized fault-tolerant control system design for unstable processes. Chem Eng Sci 58(22):5045–5054

Belcastro CM (2006) Performance Analysis on fault tolerant control system. IEEE Trans Control Syst Technol 14(5):920–925

Chen J, Cao X, Cheng P, Xiao Y, Sun Y (2010) Distributed collaborative control for industrial automation with wireless sensor and actuator networks. IEEE Trans Ind Electron 57(12):4219–4230

Chen S, Shi P, Zhang W, Zhao L (2014) Finite-time consensus on strongly convex balls of Riemannian manifolds with switching directed communication topologies. J Math Anal Appl 409(2):663–675

Chen C, Lewis FL, Xie XL, Modares H, Liu Z, Zuo S, Davoudi A (2019) Resilient adaptive and $H\infty$ controls of multi-agent systems under sensor and actuator faults. Autom 102:19–26

Chen J, Patton RJ (1999) Robust model-based fault diagnosis for dynamic systems. Kluwer Academic, Boston

Cruz D, McClintock J, Perteet B, Orqueda O, Cao Y, Fierro R (2007) Decentralized cooperative control-a multivehicle platform for research in networked embedded systems. IEEE Control Syst Mag 27(3):58–78

Defoort M, Floquet T, Kokosy A, Perruquetti W (2008) Sliding-mode formation control for cooperative autonomous mobile robots. IEEE Trans Ind Electron 55(11):3944–3953

Deng C, Gao W, Che W (2020) Distributed adaptive fault-tolerant output regulation of heterogeneous multi-agent systems with coupling uncertainties and actuator faults. IEEE/CAA J Autom Sin 7(4):1098–1106

Dong JX, Yang G (2008) Robust static output feedback control for linear discrete-time systems with time-varying uncertainties. Syst & Control Lett 57(2):123–131

Du D, Jiang B, Shi P (2011) Active fault-tolerant control for switched systems with time delay. Int J Adapt Control Signal Process 25(5):466–480

Du H, Li S, Shi P (2012) Robust consensus algorithm for second-order multi-agent systems with external disturbances. Int J Control 85(12):1913–1928

Du D, Jiang B, Shi P (2015) Fault tolerant control for switched linear systems. Springer-Verlag, Cham, Heidelberg

Frank PM (1994) Online fault detection in uncertain nonlinear systems using diagnostic observers: a survey. Int J Syst Sci 25(12):2129–2154

Frank PM (1996) Analytical and qualitative model-based fault diagnosis-a survey and some new results. Eur J Control 2(1):6–28

Frank PM, Seliger R (1990) Fault diagnosis in dynamic systems using analytical and knowledge-based redundancy-A survey and some new results. Autom 26(3):459–474

Fridman E, Shaked U (2002) An improved stabilization method for linear time-delay systems. IEEE Trans Autom Control 47(11):1931–1937

Fu Q, Shen QK, Jia Z (2020) Cooperative adaptive tracking control for uncertain nonlinear multi-agent systems with signal transmission faults. Circuit Syst Signal Process 39:1335–1352

Gao F, Zhang H (2006) Stability of time-delay fault tolerant control systems with Markovian parameters. J Beijing Univ Aeronaut Astronaut 32(5):566–570

Gertler JJ (1988) Survey of model-based failure detection and isolation in complex plants. IEEE Control Syst Mag 8(6):3–11

Gong JY, Jiang B, Shen QK (2019) Distributed-observer-based faults tolerant control design for nonlinear multi-agent systems. Int J Control Autom Syst 17(12):3149–3157

Gong JY, Jiang B, Shen QK (2019) Adaptive fault-tolerant neural control for large-scale systems with actuator faults. Int J Control Autom Syst 7(6):1421–1431

Gong J, Jiang B, Shen QK (2019) Distributed-observer-based faults tolerant control design for nonlinear multi-agent systems. Int J Control Autom Syst 17(12):3149–3157

Gong J, Jiang B, Shen QK (2020) Distributed adaptive output-feedback fault tolerant control for nonlinear systems with sensor faults. J Intell & Fuzzy Syst 38(4):4173–4190

Guo L, Nian X, Pan H (2013) Leader-following consensus of multi-agent systems with switching topologies and time-delays. J Control Theory Appl 11(2):306–310

He X, Wang ZD, Zhou DH (2009) Robust fault detection for networked systems with communication delay and data missing. Autom 45(11):2634–2639

Hong Y, Chen G, Bushnell L (2008) Distributed observers design for leader-following control of multi-agent networks. Autom 44(3):846–850

Hu Q, Xiao B (2011) Fault-tolerant sliding mode attitude control for flexible spacecraft under loss of actuator effectiveness. Nonlinear Dyn 64(1–2):13–23

Hu Q, Zhang Y, Huo X, Xiao B (2011) Adaptive integral-type sliding mode control for spacecraft attitude maneuvering under actuator stuck failures. Chin J Aeronaut 24(1):32–45

Hu Y, Cui F, Tu X, Li F (2021) Bayesian estimation of instantaneous speed for rotating machinery fault diagnosis. IEEE Trans Ind Electron 68(9):8842–8852

Huang CY, Stengel RF (1990) Re-structurable control using proportional-integral implicit model following. J Guid Control Dyn 13(2):303–309

Hung M-H, Chen C-F, Huang H-C, Yang H-C, Cheng F-T (2012) Modeling and optimization of energy consumption in cooperative multi-robot systems. IEEE Trans Autom Sci Eng 9(2):423–428

Jafari H, Ramezani A, Forouzanfar M (2022) Parameter-dependent Lyapunov function based fault estimation and fault-tolerant control for LPV systems. Int J Syst Sci 53(11):2374–2389

Jiang B, Staroswiecki M, Cocquempot V (2006) Fault accommodation for nonlinear dynamic systems. IEEE Trans Autom Control 51(9):1578–1583

Jiao H, Shen QK (2020) Distributed synchronization control with signal transmission delay and its to genetic regulatory networks. IEEE Access 8:125234–125243

Jin X, Yang G (2009) Robust adaptive fault-tolerant compensation control with actuator failures and bounded disturbances. Acta Autom Sin 35(3):305–309

Kabamba PT, Yang C (1991) Simultaneous controller design for linear time-invariant systems. IEEE Trans Autom Control 36(1):106–111

Kaminer I, Pascoal AM, Khargonekar PP, Coleman EE (1995) A velocity algorithm for the implementation of gain-scheduled controllers. Autom 31(8):1185–1191

Kwong WA, Passino KM, Laukonen EG, Yurkovich S (1995) Expert supervision of fuzzy learning systems for fault tolerant aircraft control. Proc IEEE 83(3):466–483

Li S, Tao G (2009) Feedback based adaptive compensation of control system sensor uncertainties. Autom 45(2):393–404

Li S, Du H, Shi P (2014) Distributed attitude control for multiple spacecraft with communication delays. IEEE Trans Aerosp Electron Syst 50(3):1765–1773

Liu P, Zhou D (2003) Robust fault tolerant control of uncertain time-delay systems. Prog Nat Sci 13(3):464–469

Liu YH, Wang ZD, Wang W (2011) Reliable H∞ filtering for discrete time-delay systems with randomly occurred nonlinearities via delay-partitioning method. Signal Process 91:713–727

Mahmoud MM, Jiang J, Zhang Y (2003) Active fault tolerant control systems. Springer-Verlag, New York

Mathiyalagan K, Park JH, Sakthivel R, Marshal Anthoni S (2014) Robust mixed H∞ and passive filtering for networked Markov jump systems with impulses. Signal Process 101:162–173

Ma Y, Zhang K, Jiang B (2023) Prescribed-time fault-tolerant control for fully actuated heterogeneous multiagent systems: a hierarchical design approach. IEEE Trans Aerosp Electron Syst. https://doi.org/10.1109/TAES.2023.3280895

Moerder DD (1989) Application of pre-computed laws in a reconfigurable aircraft flight control system. J Guid Control Dyn 12(3):325–333

Morari M (1985) Robust stability of systems with integral control. IEEE Trans Autom Control 30(4):574–588
MorseW D, Ossman KA (1990) Model-following reconfigurable flight control systems for the AFTI/F-16. J Guid Control Dyn 13(6):969–976
Murray R (2007) Recent research in cooperative control of multivehicle systems. ASME J Dyn Syst Meas Control 129(5):571–583
Napolitanob MR, Swaim RL (1991) New technique for aircraft flight control reconfiguration. J Guid Control Dyn 14(1):184–190
Nguyen MH, Tran DQ (2013) A combination trust model for multi-agent systems. Int J Innov Comput Inf Control 9(6):2405–2420
Ni W, Cheng D (2010) Leader-following consensus of multi-agent systems under fixed and switching topologies. Syst & Control Lett 59(3–4):209–217
Olfati-Saber R, Alex Fax J, Murray R (2007) Consensus and cooperation in networked multi-agent systems. Proc IEEE 95(1):215–233
Ranmamurthi K, Agogino AM (1993) Real-time expert system for fault tolerant supervisory control. J Dyn Syst Meas Control 115(3):219–227
Ren W, Moore KL, Chen Y (2007) High-order and model reference consensus algorithms in cooperative control of multivehicle systems. ASME J Dyn Syst Meas Control 129(5):678–688
Ren W, Beard R, Atkins E (2007) Information consensus in multivehicle cooperative control. IEEE Control Syst Mag 27(2):71–82
Ren W, Cao C (2011) Distributed coordination of multi-agent networks: emergent problems, models, and issues. Springer-Verlag, London
Richard P (2003) Time-delay systems: an overview of some recent advances and open problems. Autom 39(10):1667–1694
Saeks R, Murray J (1982) Fractional representation, algebraic geometry, and the simultaneous stabilization problem. IEEE Trans Autom Control 24(4):895–903
Saljak DD (1980) Reliable control using multiple control systems. Int J Control 31(2):303–329
Sedziwy A (2014) Effective graph representation supporting multi-agent distributed computing. Int J Innov Comput Inf Control 10(1):101–113
Shen QK, Shi P (2015) Distributed command filtered backstepping consensus tracking control of nonlinear multiple-agent systems in strict-feedback form. Autom 53:120–124
Shen QK, Shi P (2016) Output consensus control of multiagent systems with unknown nonlinear dead zone. IEEE Trans Syst Man Cybern Syst 46(10):1329–1337
Shen QK, Zhang TP, Zhou CY (2008) Decentralized adaptive fuzzy control of time-delayed interconnected systems with unknown backlash-like hysteresis. J Syst Eng Electron 19(6):1235–1242
Shen QK, Jiang B, Cocquempot V (2012) Fault diagnosis and estimation for near-space hypersonic vehicle with sensor faults. Proc Inst Mech Eng Part I J Syst Control Eng 226(3):302–313
Shen QK, Jiang B, Cocquempot V (2012) Adaptive fault tolerant synchronization with unknown propagation delays and actuator faults. Int J Control Autom Syst 10(5):883–889
Shen QK, Jiang B, Cocquempot V (2013) Fuzzy logic system-based adaptive fault tolerant control for near space vehicle attitude dynamics with actuator faults. IEEE Trans Fuzzy Syst 21(2):289–300
Shen QK, Jiang B, Cocquempot V (2013) Adaptive fault-tolerant backstepping control against actuator gain faults and its applications to an aircraft longitudinal motion dynamics. International Journal of Robust and Nonlinear Control 20(10):448–459
Shen QK, Jiang B, Shi P (2014) Cooperative adaptive fuzzy tracking control for networked unknown nonlinear multi-agent systems with time-varying actuator faults. IEEE Trans Fuzzy Syst 22(3):494–504
Shen QK, Shi P, Shi Y, Zhang J (2017) Adaptive output consensus with saturation and dead-zone and its application. IEEE Trans Ind Electron 64(6):5025–5034
Shen QK, Shi P, Zhu J, Zhang L (2019) Adaptive consensus control of leader-following systems with transmission nonlinearities. Int J Control 92(2):317–328

Shen QK, Shi Y, Jia R, Shi P (2021) Design on type-2 fuzzy-based distributed supervisory control with backlash-like hysteresis. IEEE Trans Fuzzy Syst 29(2):252–261
Shen Q, Jiang B, Shi P (2014) Novel neural networks-based fault tolerant control scheme with fault alarm. 44(11):2190–2201
Shin J-Y, Wu NE, Belcastro CM (2004) Adaptive linear parameter varying control synthesis for actuator failure. J Guid Control Dyn 27(5):787–794
Song Q, Hu WJ, Yin L, Soh YC (2002) Robust adaptive dead zone technology for fault-tolerant control of robot manipulators. J Intell Robot Syst 33(1):113–137
Srichande R, Walker BK (1993) Stochastic stability analysis for continuous-time fault tolerant control systems. Int J Control 57(3):433–452
Szupiluk R, Wojewnik P, Zabkowski T (2014) Smooth component analysis and MSE decomposition for ensemble methods in multi-agent environment. Int J Innov Comput Inf Control 10(4):1435–1445
Tao G, Chen S, Joshi SM (2002) An adaptive actuator failure compensation controller using output feedback. IEEE Trans Autom Control 47(3):506–511
Vidyasagar M, Viswanadham N (1985) Reliable stabilization using a multi-controller configuration. Autom 21(4):599–602
Wang W, Slotine J (2006) A theoretical study of different leader roles in networks. IEEE Trans Autom Control 51(7):1156–1161
Wang W, Wen C (2010) Adaptive actuator failure compensation control of uncertain nonlinear systems with guaranteed transient performance. Autom 46(12):2082–2091
Wang W, Wen C (2011) Adaptive compensation for infinite number of actuator failures or faults. Autom 47(10):2197–2210
Wang X, Yadav V, Balakrishnan S (2007) Cooperative UAV formation flying with obstacle/collision avoidance. IEEE Trans Control Syst Technol 15(4):672–679
Wang Y, Zhou D, Qin SJ, Wang H (2008) Active fault-tolerant control for a class of nonlinear systems with sensor faults. Int J Control Autom Syst 6(3):339–350
Wang D, Shi P, Wang W (2013) Robust filtering and fault detection of switched delay systems. Springer-Verlag, Berlin, Heidelberg
Wu H (2004) Reliable LQ fuzzy control for continuous-time nonlinear systems with actuator faults. IEEE Trans Syst Man Cybern Part B Cybern 34(4):1743–1752
Wu NE, Zhou K, Salomon G (2000) Control reconfigurability of linear time-invariant systems. Autom 36(12):1767–1771
Wu M, He Y, She J (2010) Stability analysis and robust control of time-delay systems. Science Press Beijing and Springer-Verlag, Berlin, Heidelberg
Xu XL, Shen QK (2022) Fault detection and isolation observer design for a class of high-order dynamic systems. ICIC Express Lett 17(9):1–8
Yang H, Jiang B, Cocquempot V (2010) Fault tolerant control design for hybrid systems. Springer-Verlag, Berlin, Heidelberg
You F, Tian Z, Shi S (2006) Senor fault diagnosis of time-delay systems based on adaptive observer. J Harbin Inst Technol (New Ser) 13(5):621–625
Yu CC, Fan MKH (1990) Decentralized integral controllability and D-stability. Chem Eng Sci 45(11):3299–3309
Yu H, Xia X (2012) Adaptive consensus of multi-agents in networks with jointly connected topologies. Autom 48:1783–1790
Zhang WZ, Bao J, Lee PL (2002) Decentralized unconditional stability conditions based on the passivity theorem for multi-loop control systems. Ind & Eng Chem Res 41(6):1569–1578
Zhang T, Guay M (2003) Adaptive control for a class of second-order nonlinear systems with unknown input nonlinearities. IEEE Trans Syst Man Cybern Part B Cybern 33(1):143–149
Zhang H, Lewis FL (2012) Adaptive cooperative tracking control of higher-order nonlinear systems with unknown dynamics. Autom 48(7):1432–1439
Zhang X, Polycarpou MM, Parisini T (2002) A robust detection and isolation scheme for abrupt and incipient fault in nonlinear systems. IEEE Trans Autom Control 47(4):576–593

Zhang X, Parisini T, Polycarpou MM (2004) Adaptive fault-tolerant control of nonlinear uncertain systems: an information-based diagnostic approach. IEEE Trans Autom Control 49(8):1259–1274
Zhang K, Jiang B, Cocquempot V (2009) Fast adaptive fault estimation and accommodation for nonlinear time-varying delay systems. Asian J Control 11(6):643–652
Zhang H, Lewis FL, Das A (2011) Optimal design for synchronization of cooperative systems: state feedback, observer and output feedback. IEEE Trans Autom Control 56(8):1948–1952
Zhang H, Lewis FL, Qu Z (2012) Lyapunov, adaptive, and optimal design techniques for cooperative systems on directed communication graphs. IEEE Trans Ind Electron 59(7):3026–3041
Zhang K, Jiang B, Ding SX, Zhou D (2020) Robust asymptotic fault estimation of discrete-time interconnected systems with sensor faults. IEEE Trans Cybern 52(3):1691–1700
Zhao H, Zhong M, Zhang M (2010) H∞ fault detection for linear discrete time-varying systems with delayed state. IET Theory Appl 4(11):2303–2314
Zhao H, Park JH, Zhang Y, Shen H (2014) Distributed output feedback consensus of discrete-time multi-agent systems. Neurocomputing 138:86–91
Zheng X, Shen QK (2023) Adaptive fault-tolerant control for high-order nonlinear systems with supervisory controllers and command filters. Int J Control Autom Syst 21(1):12–19
Zhu W, Cheng D (2010) Leader-following consensus of second-order agents with multiple time-varying delays. Autom 46(12):1994–1999

Chapter 2
Distributed Cooperative Neural Fault-Tolerant Sliding Mode Control Against Actuator Faults

2.1 Introduction

Sliding mode control (SMC), as a powerful control technique, has been widely applied in system control designing Defoort et al. (2008). At the first stage, the technique is used for single-agent systems Ghadiri et al. (2021); Xing et al. (2019). At recent years, it has been extended to multi-agent system Ni and Shi (2021); Wang et al. (2021).

For the last two decades, for multi-agent systems (MASs) including leader-following systems (LFSs), cooperative control has attracted extensive attention from the control community, and fruitful results have been obtained Defoort et al. (2008); Cruz et al. (2007); Wang et al. (2007); Chen et al. (2010); Murray (2007); Ren and Cao (2011); Wu and Shi (2011); Cai et al. (2012); Ejaz et al. (2012). In general, for the systems, their control problems can be divided into two classes. One is known as cooperative regulator problem, and the other is known as cooperative tracking problem. In the existing literatures, the first one is also named as synchronization, rendezvous, and consensus with/without leader(s) Zhang and Lewis (2012); Gao et al. (2022). For the problem, in general, distributed control methods were constructed, respectively, for the multiple agents in communication network, such that all agents are driven eventually to an un-prescribed common value. Note that the value can be time-invariant, time-varying. In some cases, it even is a function of the initial states of the agents Ren et al. (2007). The later problem is known as pinning control, model reference consensus, synchronization to a leader, leader-following control Hong et al. (2008), or leader-following consensus Zhu and Cheng (2010); Wang and Slotine (2006); Ren et al. (2007); Hong et al. (2008); Wang and Chen (2002). For the problem, a so-called leader is considered. As a command generator, the leader agent acts and generates the desired reference trajectory. Note that, in general, the leader cannot obtains the information from the follower agents. All other follower-agents' attempt to follow the trajectory of the leader agent. In most existing works on multi-agent systems in literature, however, the first- and second-order systems

Q. Shen, *Distributed Fault-Tolerant Consensus Control of Leader-Following Systems*, Intelligent Control and Learning Systems 11,
https://doi.org/10.1007/978-981-99-7426-9_2

are considered Yu and Xia (2012). In fact, in the engineering, lots of real systems are modeled by higher order dynamics. For example, in Khalil (2002), a single link flexible joint manipulator in Khalil (2002) was represented as a nonlinear fourth-order system. Therefore, for such high-order systems distributed on communication networks, how to design cooperative controllers scheme becomes more necessary. Although there exist lots of results about higher order cooperative control, most of them focus linear systems Jiang et al. (2009); Ren et al. (2007); Wang and Cheng (2007); Zhang et al. (2011). In recent years, in Zhang and Lewis (2012, 2010); Zhang et al. (2012), the authors firstly focused first- and second-order multi-agent systems and investigated their cooperative control problems, then the results for first- and second-order systems were extended by the authors (Zhang and Lewis) to nonlinear systems. However, the above results worked only in the healthy case and cannot deal with faulty case, which motivates us for this work.

For MASs including LFSs considered in this chapter, the presence of faults occurred in the controlled systems and/or lack of complete information is one of the main problems Bokovi (2002); Huang (2012); Yordanova and Jain (2012). Furthermore, modern sciences and technologies are developing, and modern control systems, especially MASs including the leader-following systems considered in this chapter, are becoming increasingly complex, which means that the number of sensors, actuators, and other system components involved is also increasing. It is obvious that, as the number of these parts increases, so does the probability of these parts failing, which will reduce system performance or even cause system instability. The instability will further lead to catastrophic accidents. Hence, to improve system reliability and to guarantee system stability in all situations, a large number of effective fault-tolerant control (FTC) approaches have been proposed Bokovi (2002); Semsar-Kazerooni (2006, 2007); Semsar-Kazerooni and Khorasani (2008); Bokovi et al. (2005); Zhang et al. (2012).

Speaking generally, the fault-tolerant control methods can be divided into two types: the active approach and the passive one. In general, as one of robust control techniques, passive FTC requires specific condition that the faults occurred in the controlled systems should be within a fault set and the fault set must be known in advance and further fixed Bokovi (2002); Semsar-Kazerooni (2006, 2007); Semsar-Kazerooni and Khorasani (2008); Bokovi et al. (2005). Generally, for the passive fault-tolerant control strategies, there are more conservatism, which will cause other problems. Thus, how to relax or decrease the conservatism becomes urgent and important. Therefore, active FTC strategies are proposed and developed. For active FTC, its purpose is online reconstruction or reconfiguration of the controller. That is to say, the so-called fault diagnosis module is working continuously during system operation. System stability and control performance can be recovered once the faults occurred in the systems are diagnosed. Under the assumption that the followers cannot access to the information from the leaders, two adaptive control schemes were constructed and applied such that formation was able to be maintained Bokovi (2002); Semsar-Kazerooni (2006). In real formation control, it is highly likely that some nodes in the whole system cannot follow operator commands in Semsar-Kazerooni and Khorasani (2009). The causes of this problem are not only actuator failures, but also other failures. For example, the faults' communication links, the sensor faults in measurement of neighboring states, and so on. For a team, its behavior was considered in the faulty case where the actuators of some agents became faulty Semsar-Kazerooni (2007); Semsar-Kazerooni and Khorasani (2008); Bokovi et al. (2005). In Semsar-Kazerooni and Khorasani (2009, 2010), for a team including multiple unmanned vehicles, under the condition that the actuators in some vehicles in the team became faulty, its performance is analyzed. However, only a class of 2-D systems was considered in Semsar-Kazerooni and Khorasani (2009, 2010), where not only is the system model relatively simple, but nonlinear dynamics are not involved. What's worst, the control scheme proposed in Semsar-Kazerooni and Khorasani (2009, 2010), as one of passive FTC, have more conservatism, which also motivate us for this work.

In this chapter, the cooperative control is investigated for a class of nonlinear LFSs with actuator faults, and an adaptive distributed FTC scheme is proposed. Compared with the existing works literature, main contributions in this work are presented as follows:

(1) Differing from Zhu and Cheng (2010); Wang and Slotine (2006); Ren et al. (2007); Hong et al. (2008); Wang and Chen (2002); Yu and Xia (2012), all the nodes' dynamics investigated in this work are not only high order but also unknown non-identical nonlinear. Furthermore, for external disturbance, optimal approximation errors, and the leader node's nonlinear dynamics, their upper boundaries are assumed to be known in Zhang and Lewis (2012). The assumption is removed in this work.
(2) Compared with Semsar-Kazerooni (2006, 2007); Semsar-Kazerooni and Khorasani (2008); Bokovi et al. (2005); Semsar-Kazerooni and Khorasani (2009), a new distributed active FTC method is designed for a class of nonlinear leader-following systems, which means that the conservatism in Semsar-Kazerooni (2006, 2007); Semsar-Kazerooni and Khorasani (2008); Bokovi et al. (2005); Semsar-Kazerooni and Khorasani (2009) was relaxed.
(3) The FTC method presented in this work can minimize the time delay induced by fault diagnosis, and its adverse effect on system performance is reduced. Furthermore, the fault isolation module is not necessary.
(4) An asymptotic value is used only for theoretical analysis and cannot be adopted in control design, which means that the asymptotic value cannot be used as a fault detection threshold in fault detection. Unlike Jiang et al. (2010a, b), a new decision threshold is defined and used in this work, which depends on an online computable fault indicator, which further implies that the decision algorithm is more practical than ones in Jiang et al. (2010a, b).

2.2 Problem Formation

In this section, let us firstly introduce some preliminaries, which includes some notations and basic graph theory. Secondly, for a class of leader-following systems, their consensus control is formulated. In addition, neural networks are also introduced in this section.

2.2.1 Basic Graph Theory

A class of multi-agent systems will be considered here, which consist of N followers and a leader. For the leader and followers, the information exchange among the leader and followers is often modeled as a graph. Hence, we first introduce the knowledge of basic graph theory in this subsection.

A directed graph can be represented as $G = (v, E)$, where agent set v and edge (arcs) set E are defined as $v = (v_1, \ldots, v_N)$ and $E \subseteq v \times v$, respectively. Note that, if there exists an edge from node i to node j, then $(v_i, v_j) \in E$. For G, its adjacency matrix A can be described as $A = [a_{ij}] \in R^{N\times N}$, where element a_{ij} is defined as follows: (1) $a_{ij} = 0$ if there exists no an edge from node i to node j, i.e., $(v_i, v_j) \notin E$; and (2) $a_{ij} \neq 0$ if there exists no an edge from node i to node j, i.e., $(v_i, v_j) \in E$. Throughout this chapter, a_{ii} is assumed to be equal to zero, i.e.,$a_{ii} = 0$, and the topology is also assumed to be fixed, which means that A is fixed and is not variant.

In directed graph G, for the $i - th$ agent, its in-degree d_i is defined as $d_i = \sum_{j=1}^{N} a_{ij}$. Correspondingly, for the directed graph G, its in-degree matrix and Laplacian matrix, i.e., D and L, are also defined, respectively, as $D = diag\{d_1, \ldots, d_N\} \in R^{N\times N}$ and $L = [l_{ij}] = D - A \in R^{N\times N}$. For the Laplacian matrix L, $L\underline{1} = 0$ with $\underline{1} = [1, \ldots, 1]^T \in R^{N\times 1}$. In addition, for agent i, its neighbor set can be described as $N_i = \{j|(v_j, v_i) \in E\}$. Obviously, for the $i - th$ agent, if the $j - th$ agent is its neighbor, agent i can obtain the information from the $j - th$ agent. For directed graph, however, not necessarily vice versa. For undirected graph, neighbors are interrelated. From node i to node j, if there exists a sequence of successive arcs, which can be described in the form $\{(v_i, v_l), (v_l, v_k), \ldots, (v_m, v_j)\}$, then a direct path exists. In a digraph, if there exist a special node and a directed path from the special node to every other node, then graph is said to has a spanning tree. For any two different nodes v_i and v_j in the digraph, from v_i to v_j, if there always exists a direct path, then the graph is said to be strongly connected.

2.2.2 Problem Formulation

Consider a leader-following system, which contains $N(N \leq 2)$ agents with distinct dynamics and one leader. For the leader, it is labeled 0 in the graph, and its dynamics has the following form:

$$\begin{cases} \dot{z}_{0,i}(t) = z_{0,i+1}(t), \quad i = 1, \ldots, n_0 - 1 \\ \dot{z}_{0,n_0}(t) = f_0(\bar{z}_0, t), \end{cases} \tag{2.1}$$

where n_0 is a positive integer and denotes the order number; $z_{0,i} \in R$ denotes the ith state; $\bar{z}_0 = [z_{0,1}, z_{0,2}, \ldots, z_{0,n_0}]^T \in R^{n_0}$; and $f_0(\bar{z}_0, t) \in R$ denotes a function. For $\bar{z}_0 \in R^{n_0}$ and $\forall t \geq 0$, it is assumed that $f_0(0, t) = 0$. In general, it is unknown for each follower in the graph.

For follower k, its dynamics is presented as follows:

$$\begin{cases} \dot{z}_{k,i}(t) = z_{k,i+1}(t), \quad i = 1, \ldots, n_k - 1 \\ \dot{z}_{k,n_k}(t) = f_k(\bar{z}_k) + u_k(t) + d_k(\bar{z}_k, t), \quad k = 1, \ldots, N, \end{cases} \tag{2.2}$$

where order number n_k is a positive integer; $z_{k,i} \in R$ is the ith state of follower's states; state vector $\bar{z}_k = [z_{k,1}, \ldots, z_{k,n_k}]^T \in R^{n_k}$; nonlinear term $f_k(\bar{z}_k) \in R$ is an unknown continuous function, $f_k(0) = 0$; control/protocol $u_k \in R$ will be designed later; and external disturbance $d_k(z_k, t) \in R$ is bounded and unknown.

To simply the complexity in the following derivation, for the leader and followers in the graph, in this chapter, let us assume that their orders are same, namely, $n_l = n$, $l = 1, 2, \ldots, N$.

For $i = 1, \ldots, n$, let us set $z_i = [z_{1,i}, \ldots, z_{N,i}]^T \in R^N$. Thus, by using the following compact form, the followers are described as

$$\begin{cases} \dot{z}_i(t) = z_{i+1}(t), \quad i = 1, \ldots, n-1 \\ \dot{z}_n(t) = f + u(t) + d, \end{cases} \tag{2.3}$$

where

$$\bar{z} = [\bar{x}_1^T, \ldots, \bar{z}_N^T]^T, \quad f = [f_1, \ldots, f_N]^T = [f_1(\bar{z}_1), \ldots, f_N(\bar{z}_N)]^T,$$

$$d = [d_1, \ldots, d_N]^T = [d_1(\bar{z}_1, t), \ldots, d_N(\bar{z}_N, t)]^T, \quad u = [u_1, \ldots, u_N]^T.$$

For follower $k(k = 1, \ldots, N)$, let us define the ith $(i = 1, \ldots, n)$ order tracking error as follows:

$$\delta_{k,i}(t) = z_{k,i}(t) - z_{0,i}(t). \tag{2.4}$$

For $i = 1, \ldots, n$, define $\delta_i = [\delta_{1,i}, \ldots, \delta_{N,i}]^T \in R^N$. Therefore, one has

$$\delta_i = z_i - \underline{z}_{0,i}, \tag{2.5}$$

where $\underline{z}_{0,i} = [z_{0,i}, \ldots, x_{0,i}]^T \in R^N$.

Definition 2.1 Zhang and Lewis (2012): For any $l(l = 1, 2, \ldots, n)$, if there exists a compact set $\Xi^l \in R^N$ with the property that $\{0\} \subset \Xi^l$, then the tracking error δ_l is said to be cooperative uniform ultimate boundedness (CUUB). Thus, for $\forall \delta_i(t_0) \in \Xi^l$, $\exists$ a time T_l and a boundary Π^l, $\ni ||\delta_i(t)|| \leq \Pi^l$, for $\forall t > T_i + t_0$, where $T_l(\Pi^i, \delta_m(t_0))$, $m = 1, 2, \ldots, n$.

Note that, from the leader node to node k, if there exists an edge, then its weight $b_k > 0$, otherwise $b_k = 0$, where $k = 1, 2, \ldots, N$.

Now, let us define the neighborhood synchronization error (NSE) of follower k $(k = 1, \ldots, N)$ as follows:

$$e_{k,i}(t) = \sum_{j \in N_k} a_{kj}(z_{j,i} - z_{k,i}) + b_k(z_{0,i} - z_{k,i}), \tag{2.6}$$

where $b_k \geq 0$ and $i = 1, \ldots, n$.

Notice that, for distributed control design of the leader-following systems, only relative state information is utilized. That is to say, for each follower in the system, the above NSE is only obtainable information in its control design.

Define

$$B = diag\{b_1, \ldots, b_N\}$$

$$\underline{f_0} = [f_0(x_0, t), \ldots, f_0(x_0, t)]^T \in R^N,$$

$$e_i = [e_{1,i}, \ldots, e_{N,i}]^T.$$

As doing in (2.3), by using the following compact form, the tracking errors defined in (2.6) can be developed as

$$\begin{cases} \dot{e}_i(t) = e_{i+1}(t), \quad i = 1, \ldots, n-1 \\ \dot{e}_n(t) = -(B + L)(f - \underline{f_0} + d + u(t)), \end{cases} \tag{2.7}$$

where the property $L(\underline{1}) = 0$ is used.

Let $\bar{E} \subseteq \bar{v} \times \bar{v}$ and $\bar{v} = \{v_0, v_1, v_2, \ldots, v_N\}$. Then, based on the graph G and a leader node, a new graph $\bar{G}$ is defined as $\bar{G} = \{\bar{v}, \bar{E}\}$.

During operation, faults may occur in the actuators. The so-called actuator bias fault is investigated in this chapter, which is described as

$$u_k^f = u_k + f_{k,u}(t), \tag{2.8}$$

where $t \geq t_{k,f}$, unknown fault occurrence time $t_{k,f} > 0$, time-varying signal $f_{k,u}(t)$ is bound but unknown.

The control objective in this work is to propose an active FTC method such that δ_i defined in (2.4) will converge to the origin or its neighborhoods in all situations. The method is divided into the following parts. (1) In normal (fault-free) situation, a normal controller $u(t)$ is constructed to make the error dynamics δ_i stable. (2) At the same time, for fault detection module, it is working. (3) The FTC scheme is activated once actuator faults have been detected, which integrates fault estimate algorithm. The FTC scheme can guarantee to obtain the control objective (the synchronization between the leader and followers) in the faulty situation.

Now, give two lemmas and three assumptions.

Assumption 2.1 Graph $\bar{G}$ has a spanning tree, in which the leader is the root.

Assumption 2.2 $\exists$ positive constants $M_0 > 0 \in R$ and $M_{f_0} > 0 \in R$, $\ni ||z_0(t)|| \leq M_0, \forall t \geq t_0$, and $|f_0(\bar{z}_0, t)| \leq M_{f_0}$.

Assumption 2.3 For each node k, there exists an unknown constant $M_{d,k} > 0 \in R$, $M_{fu,k} > 0 \in R$ such that $|d_k(\bar{z}_k, t)| \leq M_{d,k}$, $|f_{k,u}(t)| \leq M_{fu,k}$.

Lemma 2.1 *Zhang and Lewis (2012) Define* $q = [q_1, \ldots, q_N] = (B + L)^{-1}\underline{1}$, $P = diag\{p_i\} = diag\{1/q_i\}$, $Q = P(L + B) + (L + B)^T P$, *then* $P > 0$ *and* $Q > 0$.

Lemma 2.2 *Zhang and Lewis (2012) For* $i = 1, \ldots, n$, $||\delta_i|| \leq ||e_i||/\underline{\varsigma}(L + B)$, *where* $\underline{\varsigma}(B + L)$ *is a minimum singular value of matrix* $B + L$.

2.2.3 Neural Networks

As one of universal approximations, neural networks (NNs) have been utilized widely in control design of dynamic systems, especially uncertain nonlinear systems Zhang and Lewis (2010). Radial basis function (RBF) NNs, as an special NNs, will be approximated in the unknown continuous function derived from the closed-loop system stability analysis. In the following, RBF NNs are introduced simply as follows.

For any continuous function $\phi(s) : R^q \to R$, on a compact set Ω_s, there exists RBF NNs, where $s = [s_1, \ldots, s_q]^T \in R^q$, and q is a positive integer, which denotes the input dimension of the NNs. The function $\phi(s)$ can approximated by the RBF NNs in the following form:

$$\phi(s) = W^T \zeta(\bar{s}), \tag{2.9}$$

where NNs' input vector $\bar{s} = [s^T, 1]^T \in R^{q+1}$, $W = [W_1, \ldots, W_l]^T \in R^l$ is the weight vector of the NNs, positive integer l denotes the node number of the NNs, and basis function vector $\zeta(s)$ and basis function $\zeta_m(s)$ $(m = 1, 2, \ldots, l)$ are, respectively, defined as

$$\zeta(\bar{s}) = [\zeta_1(\bar{s}), \ldots, \zeta_l(\bar{s})]^T \tag{2.10}$$

and

$$\zeta_m(\bar{s}) = \exp(-\frac{\sum_{j=1}^{q} (z_j - \nu_{mj})^2}{(d_m)^2}), \tag{2.11}$$

where the Gaussian function width $d_m > 0$, the receptive field center $\nu_{mj} \in R$, $m = 1, \ldots, l$, $j = 1, \ldots, q$.

Let us define the ideal NN weights W^* as follows:

$$W^* = \arg \min_{W \in \Omega_W} [\sup_{z \in \Omega_z} |h(z, W) \quad h(z)|], \tag{2.12}$$

where Ω_s is a compact set, which can be sufficiently large, and

$$\Omega_W = \{W : ||W|| \leq M_W\} \tag{2.13}$$

with design parameter $M_W > 0$.

Hence, the above unknown function $\phi(s)$ is described by the NNs as follows:

$$\phi(s) = W^{*T}\zeta(\bar{s}) + \varepsilon(\bar{s}), \tag{2.14}$$

where optimal approximation error $\varepsilon(\bar{s}) \in R$ is bounded.

In this chapter, for $k = 1, \ldots, N$, the unknown functions $f_k(\bar{z}_k)$ in (2.2) can be approximated by using the above neural networks. That is to say, $f_k(\bar{z}_k) = W_k^{*T}\zeta_k(\bar{z}_k) + \varepsilon_k$.

From the approximation theory of NNs, a necessary assumption is given.

Assumption 2.4 The optimal approximation error $\varepsilon(\bar{s})$, $k = 1, \ldots, N$ is bounded, namely, $|\varepsilon_k| \le M_{\varepsilon,k}$, where $M_{\varepsilon,k} > 0 \in R$ is an unknown constant.

2.3 Main Results

For the above LFSs described by (2.1) and (2.2), in this section, two SMC schemes will be constructed. Beginning with, for the kth follower, $k = 1, 2, \ldots, N$, the following sliding mode error κ_k is firstly defined,

$$\kappa_k = \left(\frac{d}{dt} + \lambda\right)^{n-1} e_{k,1}(t) = \sum_{i=1}^{n-1} c_{k,i}e_{k,i}(t) + e_{k,n}(t), \tag{2.15}$$

where $c_{k,i} = C_{n-1}^{i-1}\alpha^{n-i}$, $i = 1, \ldots, n-1$, designed parameter $\alpha_k > 0$ and $C_{n-1}^{i-1} = \frac{(n-1)\times(n-2)\times\cdots\times(n-i+1)}{(i-1)\times(i-2)\times\cdots\times 1}$.

Let $\bar{e}_k(t) = [e_{k,1}(t), \ldots, e_{k,n}(t)]^T$.

Lemma 2.3 *For follower k ($k = 1, 2, \ldots N$), consider sliding mode error κ_k presented in (2.15), then,*

1. *if $\kappa_k = 0$, $\lim_{t\to\infty}\bar{e}_k(t) = 0$;*
2. *if $|\kappa_k| \le a_k$ and $\bar{e}_k(0) \in \Omega$, $\ni \bar{e}_k(t) \in \Omega$, $\forall t \ge 0$;*
3. *if $|\kappa_k| \le a_k$ and $\bar{e}_k(0) \notin \Omega$, $\exists T_k = (n-1)/\lambda_k$, $\ni \forall t \ge T_k$, $\bar{e}_k(t) \in \Omega_{a_k}$, where $\Omega = \{\bar{e}_k(t) \,|\, |e_{k,m}| \le 2^{(m-1)}\lambda_k^{m-n}a_k$, $m = 1, \ldots, n$.*

To decrease the complexity of derivation and describe the derivation process more clearly, the design parameters in (2.15), i.e., $c_{l,m}$, $l = 1, 2, \ldots, N, = 1, 2, \ldots, n$, are set as $\lambda_n = 1$, $c_{1,m} = c_{2,m} = \cdots = c_{N,m} = \lambda_m$.

Thus, one has

$$\kappa_k = \lambda_1 e_{k,1} + \cdots + \lambda_n e_{k,n}.$$

Let $\kappa = [\kappa_1, \ldots, \kappa_N]^T$, which denotes the so-called global sliding mode error. Further, one has

$$\kappa = \lambda_1 e_1 + \cdots + \lambda_n e_n.$$

From (2.7) in the previous section, it further follows that

$$\begin{aligned}\dot{\kappa} &= \sum_{i=1}^{n} \lambda_i \dot{e}_i \\ &= \sum_{i=1}^{n-1} \lambda_i e_{i+1} + \dot{e}_n \\ &= -(B+L)(f+d+u-\underline{f_0}) + \gamma,\end{aligned} \tag{2.16}$$

where $\gamma = \sum_{i=1}^{n-1} \lambda_i e_{i+1}$.

2.3.1 Cooperative Control Without Faults and Fault Detection

In this subsection, fault-free case is considered, and an adaptive control method is proposed to achieve the synchronization between the followers and leader. Based on the stability analysis, a fault detection mechanism is proposed to determine whether the fault occurs.

Beginning with control design, let us first define the notation: $\tilde{\bullet} = \bullet - \hat{\bullet}$, which will be applied in the following control designing and stability analysis.

Choosing matrix $P = P^T > 0 \in R^{n\times n}$ as in Lemma lema2.1, let us define a Lyapunov function as follows:

$$V_\kappa = \frac{1}{2}\kappa^T P \kappa.$$

From (2.16), the time derivative of V_κ can be described as follows:

$$\begin{aligned}\dot{V}_\kappa =& -\kappa^T P(B+D)W^{*T}\zeta + \kappa^T P\gamma - \kappa^T P(B+L)u - \\ &\kappa^T P(B+D)(d+\varepsilon-\underline{f_0}) + \\ &\kappa^T PAW^{*T}\zeta + \kappa^T PA(d+\varepsilon-\underline{f_0}),\end{aligned} \tag{2.17}$$

where $\varepsilon=[\varepsilon_1,\ldots,\varepsilon_N]^T$, $W^{*T} = diag\{W_1^{*T},\ldots,W_N^{*T}\}$, $\zeta=[\zeta_1^T(\bar{z}_1),\ldots,\zeta_N^T(\bar{z}_N)]^T$.

Now, let us define control law u and adaptive law of $\hat{W}$ and $\hat{\underline{M}}_{d\varepsilon f}$ as

$$u = -\text{sgn}(\kappa^T P(D+B))\hat{\underline{M}}_{d\varepsilon f} - \hat{f} - h\kappa + (D+B)^{-1}\gamma \tag{2.18}$$

$$\dot{\hat{W}} = \text{Proj}\{\eta_W \hat{W} + \eta_1 \zeta \kappa^T P(D+B)\} \tag{2.19}$$

$$\dot{\hat{\underline{M}}}_{d\varepsilon f} = \text{Proj}\{\eta_{Md\varepsilon}\hat{\underline{M}}_{d\varepsilon f} + \eta_2\kappa^T P(B+D)\text{sgn}(\kappa^T P(B+D))\}, \tag{2.20}$$

where design parameters h, η_W, and $\eta_{Md\varepsilon}$ are positive real numbers; $\hat{W}^T = diag\{\hat{W}_m^T\}$ with $m = 1, 2, \ldots, N$, $\hat{f} = [\hat{f}_1, \ldots, \hat{f}_N]^T$, and $\hat{\underline{M}}_{d\varepsilon f} = [\hat{M}_{d\varepsilon f,1}, \ldots, \hat{M}_{d\varepsilon f,N}]^T$;

$\hat{W}_k^T$, $\hat{f}_k = \hat{W}_k^T \zeta_k(\bar{z}_k)$ and $\hat{M}_{d\varepsilon f,k}$ denote, respectively, the estimates of W_k^{*T}, $f_k(\bar{z}_k)$ and $M_{d\varepsilon f,k} = M_{d,k} + M_{\varepsilon,k} + M_{f_0}$, $k = 1, 2, \ldots, N$. For $\kappa^T P(D + B)$, its kth element is described by $P_{\kappa,k}$, $\text{sgn}(\kappa^T P(B + D)) = diag\{\text{sgn}(P_{\kappa,m})\}, m = 1, 2, \ldots, N$, the project operator is used to ensure the boundedness of $\hat{\bar{M}}_{d\varepsilon}$ and $\hat{W}$.

By using the following theorem, the detailed stability analysis is given.

Theorem 2.1 *Under Assumptions 2.1-2.4, consider the LFSs described by (2.1) and (2.2), and control law (2.18) and adaptive law (2.19, 2.20) are adopted.* ***If*** *the parameters* $\eta_1, \eta_2, r, h, \eta_{Md\varepsilon}, \eta_W, \mu_\kappa > 0 \in R$, *and* $\mu_{WM} > 0 \in R$ *are appropriately chosen to guarantee the following inequality holds:*

$$\bar{\varsigma}(P)\bar{\varsigma}(A)(\zeta^T \tilde{W}\tilde{W}^T\zeta + \bar{M}_{d\varepsilon f}^T \bar{M}_{d\varepsilon f} + \hat{\bar{M}}_{d\varepsilon f}^T \hat{\bar{M}}_{d\varepsilon f})/(4r) \le \mu_\kappa$$

$$\eta_W tr(W^{*T} W^*)/(2\eta_1) + \eta_{Md\varepsilon} \bar{M}_{d\varepsilon f}^T \bar{M}_{d\varepsilon f}/(2\eta_2) \le \mu_{WM}$$

$$h\underline{\varsigma}(Q)/2 - (4r + \frac{\bar{\lambda}^2}{(4r\underline{\lambda}^2)})\bar{\varsigma}(P)\bar{\varsigma}(A) > 0,$$

then *the following results are obtained.*

1. *In finite time, each follower synchronizes to the leader. Meanwhile,* $\delta_m(t)$ *(*$m = 1, 2, \ldots, N$*) and all the signals in the resulting closed-loop system, respectively, belong to two adjustable sets* Ω_δ *and* Ω*, which are described, respectively, as follows:*

$$\Omega_\delta = \{\delta_m \mid ||\delta_m|| \le \sqrt{N(2^{(m-1)}\lambda_k^{(m-n)} M_V)^2}\}, k = 1, \ldots, N, m = 1, \ldots, n;$$

$$\Omega = \left\{ \begin{array}{l} (\tilde{\bar{M}}_{d\varepsilon f}(t), \kappa(t), \tilde{W}(t)) \Big| \; ||\tilde{\bar{M}}_{d\varepsilon f}(t)|| \le \sqrt{\eta_2} M_V, \\ ||\tilde{W}(t)|| \le \sqrt{\eta_1} M_V, ||\kappa(t)|| \le M_V \end{array} \right\}$$

.

2. *For* $\forall t \ge t_0$, $k = 1, \ldots, N$, *the followers' states are bounded.*

Proof Let us substitute (2.18) into (2.17), then

$$\begin{aligned}
\dot{V}_\kappa = & \kappa^T P(B + D)\tilde{W}^T\zeta - h\kappa^T P(B + L)\kappa \\
& + \kappa^T PA\tilde{W}^T\zeta + \kappa^T P(B + D)\text{sgn}(\kappa^T P(B + D))\tilde{\bar{M}}_{d\varepsilon f} \\
& + \kappa^T PA(d + \varepsilon - \underline{f_0}) + \kappa^T PA(B + D)\gamma - \kappa^T PA\text{sgn}(\kappa^T P(B + D))\hat{\bar{M}}_{d\varepsilon f}.
\end{aligned}$$

Since

$$\kappa^T PA\tilde{W}^T\zeta \le \frac{\bar{\varsigma}(A)\bar{\varsigma}(P)}{4r}\zeta^T\tilde{W}\tilde{W}^T\zeta + \bar{\varsigma}(A)\bar{\varsigma}(P)r\kappa^T\kappa,$$

$$\kappa^T PA(d+\varepsilon-\underline{f_0} \le \frac{\bar{\varsigma}(A)\bar{\varsigma}(P)}{4r}\bar{M}_{d\varepsilon f}^T\bar{M}_{d\varepsilon f} + \bar{\varsigma}(A)\bar{\varsigma}(P)r\kappa^T\kappa,$$

$$\frac{\bar{\varsigma}(P)\bar{\varsigma}(A)}{4r}\hat{\bar{M}}_{d\varepsilon f}^T\hat{\bar{M}}_{d\varepsilon f} + \bar{\varsigma}(A)\bar{\varsigma}(P)r\kappa^T\kappa \ge -\kappa^T PA\mathrm{sgn}(\kappa^T P(B+D))\hat{\bar{M}}_{d\varepsilon f} \le$$

$$\bar{\varsigma}(A)\bar{\varsigma}(P)/(4r)\gamma^T\gamma + \bar{\varsigma}(A)\bar{\varsigma}(P)r\kappa^T\kappa \ge \kappa^T PA(B+D)\gamma,$$

where $\bar{M}_{d\varepsilon f} = [M_{d\varepsilon f,1},\ldots,M_{d\varepsilon f,N}]^T$ and design parameter r is a positive real number, it follows

$$\begin{aligned}\dot{V}_\kappa \le&\kappa^T P(B+D)\tilde{W}^T\zeta - h\kappa^T P(B+L)\kappa\\ &+\kappa^T P(B+D)\mathrm{sgn}(\kappa^T P(B+D))\tilde{\bar{M}}_{d\varepsilon f}\\ &+\bar{\varsigma}(A)\bar{\varsigma}(P)/(4r)(\zeta^T\tilde{W}\tilde{W}^T\zeta + 4r\bar{\varsigma}(A)\bar{\varsigma}(P)\kappa^T\kappa\\ &+\hat{\bar{M}}_{d\varepsilon f}^T\hat{\bar{M}}_{d\varepsilon f} + \bar{M}_{d\varepsilon f}^T\bar{M}_{d\varepsilon f} + \gamma^T\gamma\bar{\varsigma}(A)\bar{\varsigma}(P)/(4r).\end{aligned}$$

Notice that, for $\hat{W}_k$ and W_k^*, their boundedness can be guaranteed, respectively, by the adaptive law defined by (2.19) and Assumption 2.4. Obviously, the estimate error W, i.e., $\tilde{W} = W^* - hat W$, is also bounded. From the definition of ζ defined in the previous section, one has $\zeta_k(\bar{z}_k) \le 1$. Hence, the boundedness of $\tilde{W}$ and ζ can directly lead to the following conclusion: $\tilde{W}^T\zeta$ is bounded.

On the other hand, from Assumptions 2.2 and 2.3, one has that f_0, d_k, and ε_k are bounded, which means that $\bar{M}_{d\varepsilon f}$ is also bounded. In addition, the boundedness of $\hat{\bar{M}}_{d\varepsilon f}$ can be guaranteed by the adaptive law in (2.20). Hence, the estimate error of $\bar{M}_{d\varepsilon f}$, i.e., $\tilde{\bar{M}}_{d\varepsilon f}$ is bounded.

From the boundedness of $\hat{\bar{M}}_{d\varepsilon f}^T$, $\bar{M}_{d\varepsilon f}$ and $\tilde{W}^T\zeta$, we can obtain, if design parameter r is set to be a large enough positive real number, then the following inequality holds:

$$\frac{\bar{\varsigma}(P)\bar{\varsigma}(A)}{4r}(\zeta^T\tilde{W}\tilde{W}^T\zeta + \bar{M}_{d\varepsilon f}^T\bar{M}_{d\varepsilon f} + \hat{\bar{M}}_{d\varepsilon f}^T\hat{\bar{M}}_{d\varepsilon f}) \le \mu_\kappa,$$

where design parameter $\mu_\kappa > 0 \in R$. Because

$$\gamma_k^2 = \sum_{i=1}^{n-1}\lambda_i^2 e_{k,i+1}^2 \le \bar{\lambda}^2\kappa_k^2/\underline{\lambda}^2,$$

where $\bar{\lambda} = \max\{\lambda_1,\ldots,\lambda_n\}$, $\underline{\lambda} = \min\{\lambda_1,\ldots,\lambda_n\}$, one has

$$\gamma^T\gamma = \sum_{k=1}^{N} \gamma_k^2 \leq \sum_{k=1}^{N} \frac{\bar{\lambda}^2}{\underline{\lambda}^2}\kappa_k^2 = \frac{\bar{\lambda}^2}{\underline{\lambda}^2}\sum_{k=1}^{N} \kappa_k^2 = \frac{\bar{\lambda}^2}{\underline{\lambda}^2}\kappa^T\kappa.$$

Hence, $\dot{V}_\kappa$ can be developed as

$$\begin{aligned}\dot{V}_\kappa \leq & -h\kappa^T P(B+L)\kappa + \kappa^T P(B+D)\tilde{W}^T\zeta \\ & + \kappa^T P(B+D)\mathrm{sgn}(\kappa^T P(B+D))\tilde{\bar{M}}_{d\varepsilon f} \\ & + \frac{\bar{\lambda}^2+4r}{4r\underline{\lambda}^2}\bar{\varsigma}(A)\bar{\varsigma}(P)\kappa^T\kappa + \mu_\kappa.\end{aligned}$$

Define

$$V_W = tr\{\tilde{W}^T\tilde{W}\}/(2\eta_1) + \tilde{\bar{M}}_{d\varepsilon f}^T\tilde{\bar{M}}_{d\varepsilon f}/(2\eta_2).$$

Its time derivative, i.e., $\dot{V}_W$, can be described as

$$\dot{V}_W = -\frac{tr\{\tilde{W}^T\dot{\hat{W}}\}}{\eta_1} - \frac{\tilde{\bar{M}}_{d\varepsilon f}^T\dot{\hat{\bar{M}}}_{d\varepsilon f}}{\eta_2}. \tag{2.21}$$

Define

$$V_0 = V_\kappa + V_W.$$

Its time derivative, i.e., $\dot{V}_0$, can be described as

$$\begin{aligned}\dot{V}_0 \leq & -h\kappa^T P(B+L)\kappa + (4r + \frac{1}{4r}\frac{\bar{\lambda}^2}{\underline{\lambda}^2})\bar{\varsigma}(P)\bar{\varsigma}(A)\kappa^T\kappa \\ & + \mu_\kappa + \kappa^T P(B+D)\tilde{W}^T\zeta + \kappa^T P(B+D)\mathrm{sgn}(\kappa^T P(B+D))\tilde{\bar{M}}_{d\varepsilon f} \\ & - \frac{1}{\eta_1}tr\{\tilde{W}^T\dot{\hat{W}}\} - \frac{1}{\eta_2}\tilde{\bar{M}}_{d\varepsilon f}^T\dot{\hat{\bar{M}}}_{d\varepsilon f}.\end{aligned} \tag{2.22}$$

Substituting (2.19) and (2.20) into (2.22), it follows

$$\begin{aligned}\dot{V}_0 \leq & -h\kappa^T P(B+L)\kappa + (\frac{1}{4r}\frac{\bar{\lambda}^2}{\underline{\lambda}^2} + 4r)\bar{\varsigma}(A)\bar{\varsigma}(P)\kappa^T\kappa \\ & - \frac{\eta_W}{\eta_1}tr(\tilde{W}^T\hat{W}) - \frac{\eta_{Md\varepsilon}}{\eta_2}\tilde{\bar{M}}_{d\varepsilon f}^T\hat{\bar{M}}_{d\varepsilon f} + \mu_\kappa.\end{aligned}$$

Since

$$-\frac{\eta_W}{\eta_1}tr(\tilde{W}^T\hat{W}) \leq -\frac{\eta_W}{2\eta_1}tr(\tilde{W}^T\tilde{W}) + \frac{\eta_W}{2\eta_1}tr(W^{*T}W^*)$$

$$-\frac{\eta_{Md\varepsilon}}{\eta_2}\tilde{\bar{M}}_{d\varepsilon f}^T\hat{\bar{M}}_{d\varepsilon f} \leq -\frac{\eta_{Md\varepsilon}}{2\eta_2}\tilde{\bar{M}}_{d\varepsilon f}^T\tilde{\bar{M}}_{d\varepsilon f} + \frac{\eta_{Md\varepsilon}}{2\eta_2}\bar{M}_{d\varepsilon f}^T\bar{M}_{d\varepsilon f},$$

one has

$$\dot{V}_0 \leq -\frac{1}{2}h\kappa^T Q\kappa + (4r + \frac{1}{4r}\frac{\bar{\lambda}^2}{\underline{\lambda}^2})\bar{\varsigma}(P)\bar{\varsigma}(A)\kappa^T\kappa -$$
$$\frac{\eta_W}{2\eta_1}tr(\tilde{W}^T\tilde{W}) - \frac{\eta_{Md\varepsilon}}{2\eta_2}\tilde{\bar{M}}^T_{d\varepsilon f}\tilde{\bar{M}}_{d\varepsilon f} +$$
$$\frac{\eta_W}{2\eta_1}tr(W^{*T}W^*) + \frac{\eta_{Md\varepsilon}}{2\eta_2}\bar{M}^T_{d\varepsilon f}\bar{M}_{d\varepsilon f} + \mu_\kappa.$$

From the previous analysis, it is easy to find that $\bar{M}_{d\varepsilon f}$ and W^* are bounded. Thus, design parameters $\eta_1, \eta_2, \eta_{Md\varepsilon}, \eta_W$, and $\mu_{WM} > 0 \in R$are appropriately chosen such that

$$\frac{\eta_W}{2\eta_1}tr(W^{*T}W^*) + \frac{\eta_{Md\varepsilon}}{2\eta_2}\bar{M}^T_{d\varepsilon f}\bar{M}_{d\varepsilon f} \leq \mu_{WM}.$$

Let $\mu_0 = \mu_{WM} + \mu_\kappa$, then

$$\dot{V}_0 \leq -\lambda_0 V_0 + \mu_0.$$

Further, one has

$$0 \leq V_0(t) \leq \frac{\mu_0}{\lambda_0} + (V_0(t_0) - \frac{\mu_0}{\lambda_0})e^{-\lambda_0(t-t_0)},$$

where

$$\lambda_0 = \min\{\frac{\frac{1}{2}h\underline{\varsigma}(Q) - (4r + \frac{1}{4r}\frac{\bar{\lambda}^2}{\underline{\lambda}^2})\bar{\varsigma}(P)\bar{\varsigma}(A)}{\bar{\varsigma}(P)}, \frac{\eta_W}{2\eta_1}, \frac{\eta_{Md\varepsilon}}{2\eta_2}\}.$$

Let $V_0(T_m) = \frac{1}{2}\underline{\varsigma}(P)M_V^2$, design parameter $M_V > 0 \in R$. Thus, based on the above equality, it further follows

$$T_m = \frac{\ln\frac{V_0(T_m) - \frac{\mu_0}{\lambda_0}}{(V_0(t_0) - \frac{\mu_0}{\lambda_0})}}{-\lambda_0} + t_0 = \frac{ln\frac{\frac{1}{2}\underline{\varsigma}(P)M_V^2 - \frac{\mu_0}{\lambda_0}}{(V_0(t_0) - \frac{\mu_0}{\lambda_0})}}{-\lambda_0} + t_0. \tag{2.23}$$

From (2.23), it easy to obtain that, for $t \geq T_m$, $V_0(t) \leq \frac{\underline{\varsigma}(P)M_V^2}{2}$. Because $\underline{\varsigma}(P)||\kappa(t)||^2 \leq 2V_\kappa(t) \leq 2V_0(t)$, one has

$$||\kappa(t)|| \leq \sqrt{2V_0(t)/\underline{\varsigma}(P)}.$$

Further, it follows that, for $\forall t \geq T_m$,

$$||\kappa(t)|| \leq M_V.$$

Similar to the analysis of $\kappa(t)$, it further follows, for $\forall t \geq T_m$,

$$||\tilde{\tilde{M}}_{d\varepsilon f}(t)|| \leq \sqrt{\eta_2} M_V$$

and

$$||\tilde{W}(t)|| \leq \sqrt{\eta_1} M_V.$$

On the other hand, for $\forall t \geq T_m$, all signals are within a small set, i.e., Ω, which is presented as

$$\Omega = \left\{ \begin{array}{l} (\kappa(t), \tilde{W}(t), \tilde{\tilde{M}}_{d\varepsilon f}(t)) \Big| \, ||\kappa(t)|| \leq M_V, \\ ||\tilde{W}(t)|| \leq \sqrt{\eta_1} M_V, ||\tilde{\tilde{M}}_{d\varepsilon f}(t)|| \leq \sqrt{\eta_2} M_V \end{array} \right\}.$$

It is easy to find that, by setting the parameters r, η_W, $\eta_{Md\varepsilon}$, η_1 and η_2, Ω is adjusted.

Because $\kappa = [\kappa_1, \ldots, \kappa_N]^T$ and $||\kappa(t)|| \leq M_V$, one can obtain that $|\kappa_k(t)| \leq M_V$. It further follows from Lemma lema2.3 that, for $m = 1, \ldots, n$ and $k = 1, \ldots, N$, $|e_{k,m}(t)| \leq 2^{(m-1)} \cdot \lambda_k^{(m-n)} M_V$. And because $e_m = [e_{1,m}, \ldots, e_{N,m}]^T \in R^N$, it can be obtained that $||e_m(t)|| \leq \sqrt{N(M_V \times 2^{(m-1)} \lambda_k^{(m-n)})^2}$.

Finally, it follows from Lemma lema2.2 that, for $k = 1, \ldots, N$ and $m = 1, \ldots, n$,

$$||\delta_m(t)|| \leq \frac{\sqrt{N(M_V \times 2^{(m-1)} \lambda_k^{(m-n)})^2}}{\underline{\varsigma}(L + B)}.$$

From Assumption 2.2, the leader's state is bounded. And from the above analysis, the tracking error δ_k is bounded, $k = 1, 2, \ldots, N$. Hence, it can be obtained that, for $k = 1, \ldots, N$, the state $\bar{z}_k(t)$ is bounded. The proof is completed here.

As stated in the previous sections, in engineering applications, failures occurred in the actuators of the controlled systems. Thus, fault detection residual J is set as $\Delta = ||\kappa(t)||$, which will be used in fault detection. It is obvious to find that, in healthy case, the following condition holds:

$$\text{For} \quad \forall t \geq T_m, \quad \Delta = ||\kappa(t)|| \leq M_V.$$

In the chapter, let us assume that, at time $t \geq T_m$, the faults occur. The detailed mechanism is described as follows:

$$\begin{cases} \Delta > M_V & \text{faults have occurred} \\ \Delta \leq M_V & \text{no fault occurred.} \end{cases} \tag{2.24}$$

In the following subsection, a novel FTC scheme is proposed, which does not require the additional fault isolation module.

2.3.2 Distributed FTC Against Actuator Faults

In this subsection, it is assumed that follower k is faulty, $k = 1, \ldots, N$. Based on the fault model (2.8), its dynamics is re-described in the following form:

$$\begin{cases} \dot{z}_{k,i}(t) = z_{k,i+1}(t) \\ \dot{z}_{k,n}(t) = d_k(\bar{z}_k, t) + f_{k,u}(t) + u_k(t) + f_k(\bar{z}_k), \end{cases} \tag{2.25}$$

where $i = 1, \ldots, n-1$.

Let

$$f_u(t) = [f_{1,u}(t), f_{2,u}(t), , \ldots, f_{N,u}(t)]^T.$$

In the following analysis, $f_{k,u}(t)$ and $f_u(t)$ will be denoted respectively by their abbreviations, i.e., $f_{k,u}$, f_u, $k = 1, 2, \ldots, N$. Then, similar to (2.3), for the followers in the graph, in the compact form, their dynamics is re-written as

$$\begin{cases} \dot{z}_i(t) = z_{i+1}(t), \\ \dot{z}_n(t) = f + d + u(t) + f_u \end{cases} \tag{2.26}$$

where $i = 1, \ldots, n-1$.

Correspondingly, the tracking error dynamics defined by (2.7) in the previous section can be re-descried in the following form:

$$\begin{cases} \dot{e}_i(t) = e_{i+1}(t), i = 1, 2, \ldots, n-1 \\ \dot{e}_n(t) = -(B + L)(f + d + f_u + u(t) - \underline{f_0}) \end{cases} \tag{2.27}$$

Similar to doing in the previous subsection, Lyapunov function V_κ is defined as $V_\kappa = \frac{1}{2}\kappa^T P \kappa$. Differentiating V_κ, it follows from (2.26) and (2.27) that

$$\begin{aligned} \dot{V}_\kappa &= \kappa^T P \dot{\kappa} \\ &= \kappa^T P(\gamma - (B + L)(f + d + f_u + u - \underline{f_0})). \end{aligned} \tag{2.28}$$

Now, the following adaptive and control laws are defined respectively such that the faults can be compensated,

$$u = (B + D)^{-1}\gamma - \text{sgn}(\kappa^T P(B + D))\hat{\bar{M}}_{d\varepsilon f} - \hat{f}_u - \hat{f} - h\kappa, \tag{2.29}$$

where $\hat{f}_u$ denotes the estimate of f_u.

$$\dot{\hat{W}} = \text{Proj}\{\eta_1 \kappa(z)\kappa^T P(B + D) + \eta_W \hat{W}\} \tag{2.30}$$

$$\dot{\hat{\bar{M}}}_{d\varepsilon f} = \text{Proj}\{\eta_{Md\varepsilon}\hat{\bar{M}}_{d\varepsilon f} + \eta_2 \kappa^T P(B + D)\text{sgn}(\kappa^T P(B + D))\} \tag{2.31}$$

and

$$\dot{\hat{f}}_u = \text{Proj}\{\eta_{fu}\hat{f}_u + \eta_3\kappa^T P(B+D).\} \tag{2.32}$$

In this chapter, the faults are assumed to be bounded. Similarly, as doing in the previous subsection, by choosing appropriately the design parameter $r > 0$, one obtains that

$$\frac{\bar{\varsigma}(P)\bar{\varsigma}(A)}{4r} \times (\zeta^T\tilde{W}\tilde{W}^T\zeta + \hat{\tilde{M}}^T_{d\varepsilon f}\hat{\tilde{M}}_{d\varepsilon f} + \bar{M}^T_{d\varepsilon f}\bar{M}_{d\varepsilon f} + \tilde{f}^T_u\tilde{f}_u) \le \mu'_\kappa,$$

where μ'_κ is a design parameter.

Theorem 2.2 *Under Assumptions 2.1–2.4, consider the faulty leader-following system (2.26) and (2.1) with the fault (2.8). Adopting the adaptive laws and control input (2.29)–(2.32), if the parameters $\mu'_\kappa > 0$, $\mu'_{WM} > 0$, η_W, h, r, $\eta_{Md\varepsilon}$, and η_l, $l = 1, 2, 3$ are chosen appropriately to ensure the following inequalities hold:*

$$\frac{\bar{\varsigma}(P)\bar{\varsigma}(A)}{4r}(\zeta^T\tilde{W}\tilde{W}^T\zeta + \tilde{f}^T_u\tilde{f}_u + \bar{M}^T_{d\varepsilon f}\bar{M}_{d\varepsilon f} + \hat{\tilde{M}}^T_{d\varepsilon f}\hat{\tilde{M}}_{d\varepsilon f}) \le \mu'_\kappa$$

$$\frac{\eta_W}{2\eta_1}tr(W^{*T}W^*) + \frac{\eta_{Md\varepsilon}}{2\eta_2}\bar{M}^T_{d\varepsilon f}\bar{M}_{d\varepsilon f} + +\frac{\eta_{fu}}{2\eta_3}f^T_u f_u \le \mu'_{WM}$$

$$\frac{1}{2}h\underline{\varsigma}(Q) - (4r + \frac{1}{4r}\frac{\bar{\lambda}^2}{\underline{\lambda}^2})\bar{\varsigma}(A)\bar{\varsigma}(P) > 0,$$

then the following results can be obtained:

1. *In finite time, each follower synchronizes to the leader, and tracking error $\delta_m(t)$ $(m = 1, 2, \ldots, N)$ and all the signals in the resulting closed-loop system belong, respectively, to two adjustable sets Ω'_δ and Ω', which are defined, respectively, as follows: for $k = 1, \ldots, N$,*

$$\Omega' = \left\{ \begin{array}{l} (\kappa(t), \tilde{W}(t), \tilde{\bar{M}}_{d\varepsilon f}(t), \tilde{f}_u(t))\Big| \ ||\tilde{W}(t)|| \le \sqrt{\eta_1}M_V, ||\tilde{f}_u(t)|| \le \sqrt{\eta_3}M_V, \\ , ||\tilde{\bar{M}}_{d\varepsilon f}(t)|| \le \sqrt{\eta_2}M_V, ||\kappa(t)|| \le M_V \end{array} \right\}$$

$$\Omega'_\delta = \{\delta_m \mid ||\delta_m|| \le \sqrt{N \times M_V(2^{(m-1)}\lambda_k^{(m-n)})^2}\},$$

$$m = 1, \ldots, n.$$

2. *For $\forall t \ge t_0$, $k = 1, \ldots, N$, the followers' states are bounded.*

Proof Substituting control law (2.29) into (2.28), it yields that

$$\dot{V}_\kappa \leq -h\kappa^T P(B+L)\kappa + (4r + \frac{1}{4r}\frac{\bar{\lambda}^2}{\underline{\lambda}^2})\bar{\varsigma}(A)\bar{\varsigma}(P)\kappa^T\kappa +$$
$$\kappa^T P(B+D)\text{sgn}(\kappa^T P(B+D))\tilde{\bar{M}}_{d\varepsilon f} + \mu'_\kappa + \kappa^T P(B+D)\tilde{W}^T\zeta.$$

Define

$$V_W = \frac{1}{2\eta_1}tr\{\tilde{W}^T\tilde{W}\} + \frac{1}{2\eta_2}\tilde{\bar{M}}^T_{d\varepsilon f}\tilde{\bar{M}}_{d\varepsilon f} + \frac{1}{2\eta_3}\tilde{f}_u^T\tilde{f}_u.$$

Its time derivative, i.e., $\dot{V}_W$, is developed as

$$\dot{V}_W = -\frac{1}{\eta_1}tr\{\tilde{W}^T\dot{\hat{W}}\} - \frac{1}{\eta_2}\tilde{\bar{M}}^T_{d\varepsilon f}\dot{\hat{\bar{M}}}_{d\varepsilon f} - \frac{1}{\eta_3}\tilde{f}_u^T\dot{\hat{f}}_u.$$

Define

$$V_1 = V_\kappa + V_W.$$

Its time derivative, i.e., $\dot{V}_0$, is developed as

$$\begin{aligned}\dot{V}_0 =& -\kappa^T P\kappa + tr\{\tilde{W}^T\zeta\kappa^T P(B+L)\} + \\ &\kappa^T P(B+L)\text{sgn}(\kappa^T P(B+L))\tilde{\bar{M}}_{d\varepsilon f} - \\ &tr\{\tilde{W}^T\dot{\hat{W}}\}/\eta_1 - \tilde{\bar{M}}^T_{d\varepsilon f}\dot{\hat{\bar{M}}}_{d\varepsilon f}/\eta_2 - \tilde{f}_u^T\dot{\hat{f}}_u/\eta_3.\end{aligned} \tag{2.33}$$

From Young's inequality, one has

$$-\frac{\eta_{fu}}{\eta_3}\tilde{f}_u^T\hat{f}_u \leq -\frac{\eta_{fu}}{2\eta_3}\tilde{f}_u^T\tilde{f}_u + \frac{\eta_{fu}}{2\eta_3}\bar{M}^T_{fu}\bar{M}_{fu} \tag{2.34}$$

where $\bar{M}_{fu} = [M_{fu,1}, \ldots, M_{fu,N}]^T$.

Substituting the above adaptive laws (2.30)–(2.32) into (2.33), one has

$$\dot{V}_1 \leq -\frac{1}{2}h\kappa^T Q\kappa + (4r + \frac{1}{4r}\frac{\bar{\lambda}^2}{\underline{\lambda}^2})\bar{\varsigma}(P)\varsigma(A)\kappa^T\kappa - \frac{\eta_W}{2\eta_1}tr(\tilde{W}^T\tilde{W}) -$$
$$\frac{\eta_{Md\varepsilon}}{2\eta_2}\tilde{\bar{M}}^T_{d\varepsilon f}\tilde{\bar{M}}_{d\varepsilon f} - \frac{\eta_{fu}}{2\eta_3}\tilde{f}_u^T\tilde{f}_u + \frac{\eta_W}{2\eta_1}tr(W^{*T}W^*) +$$
$$\frac{\eta_{Md\varepsilon}}{2\eta_2}\bar{M}^T_{d\varepsilon f}\bar{M}_{d\varepsilon f} + \mu'_\kappa + \frac{\eta_{fu}}{2\eta_3}\bar{M}^T_{fu}\bar{M}_{fu}.$$

As doing in the previous subsection, by choosing appropriately the design parameters $\mu'_{WM} > 0$, $\eta_{Md\varepsilon}$, η_W, and η_l, $l = 1, 2$, the following inequality holds:

$$\frac{\eta_W}{2\eta_1} tr(W^{*T} W^*) + \frac{\eta_{Md\varepsilon}}{2\eta_2} \bar{M}_{d\varepsilon f}^T \bar{M}_{d\varepsilon f} + \frac{\eta_{fu}}{2\eta_3} \bar{M}_{fu}^T \bar{M}_{fu} \leq \mu'_{WM}.$$

Let $\mu_1 = \mu'_{WM} + \mu'_{\kappa}$, then one has

$$\dot{V}_1 \leq -\lambda_1 V_1 + \mu_1, \tag{2.35}$$

where

$$\lambda_1 = \min\{\frac{\frac{1}{2} h \underline{\varsigma}(Q) - (4r + \frac{1}{4r}\frac{\bar{\lambda}^2}{\underline{\lambda}^2}) \bar{\varsigma}(P) \bar{\varsigma}(A)}{\bar{\varsigma}(P)}, \frac{\eta_W}{2\eta_1}, \frac{\eta_{Md\varepsilon}}{2\eta_2}, \frac{\eta_{fu}}{2\eta_3}\}.$$

Let $V_1(T_m) = \frac{1}{2}\underline{\varsigma}(P) M_V^2$, design parameter $M_V > 0 \in R$. Then, one can obtain

$$T_m = \frac{\ln \frac{V_1(T_m) - \frac{\mu_1}{\lambda_1}}{(V_1(t_0) - \frac{\mu_1}{\lambda_1})}}{-\lambda_1} + t_0 = \frac{\ln \frac{\frac{1}{2}\underline{\varsigma}(P) M_V^2 - \frac{\mu_1}{\lambda_1}}{(V_1(t_0) - \frac{\mu_1}{\lambda_1})}}{-\lambda_1} + t_0. \tag{2.36}$$

From (2.36), it can be obtained that

$$\text{For } t \geq T_m,\ V_1(t) \leq \frac{\underline{\varsigma}(P) M_V^2}{2}.$$

Because

$$\underline{\varsigma}(P) ||\kappa(t)||^2 \leq 2V_\kappa(t) \leq 2V_1(t),$$

one has

$$||\kappa(t)|| \leq \sqrt{2V_1(t)/\underline{\varsigma}(P)}.$$

Further, one obtains

$$\forall t \geq T_m,\ ||\kappa(t)|| \leq M_V.$$

It is easy to find from the above equation that, $\exists\ M_V$ and $T_m, \ni$ for $\forall t \geq T_m, ||\kappa(t)|| \leq M_V$. Similarly, $||\tilde{W}(t)|| \leq M_V \sqrt{\eta_1}$, $||\tilde{\bar{M}}_{d\varepsilon f}(t)|| \leq M_V \sqrt{\eta_2}$, $||\tilde{f}_u(t)|| \leq M_V \sqrt{\eta_3}$, for $\forall t \geq T_m$. It means that the closed-loop signals are within Ω' for $t \geq T_m$, which is described as

$$\Omega' = \left\{ \begin{array}{l} (\kappa(t), \tilde{W}(t), \tilde{\bar{M}}_{d\varepsilon f}(t), \tilde{f}_u(t)) \Big| \ ||\kappa(t)|| \leq M_V, ||\tilde{W}(t)|| \leq \\ \sqrt{\eta_1} M_V, ||\tilde{\bar{M}}_{d\varepsilon f}(t)|| \leq \sqrt{\eta_2} M_V, ||\tilde{f}_u(t)|| \leq \sqrt{\eta_3} M_V \end{array} \right\}.$$

Obviously, by choosing the parameters $\eta_{Md\varepsilon}, \eta_W, r, h, \eta_l, l = 1, 2, 3$, Ω' is adjusted. Similar to Theorem 2.1, the following result can be obtained:

$$||\delta_m(t)|| \leq \frac{\sqrt{N (2^{(m-1)} \lambda_k^{(m-n)} M_V)^2}}{\underline{\varsigma}(B + L)},\ k = 1, 2, \ldots, N,\ m = 1, 2, \ldots, n.$$

Similar to the analysis in Theorem 2.1, for $k = 1, \ldots, N\kappa(t)$ and $\kappa_k(t)$ are CUUB. From Lemma lema2.3, it is further obtained that, for $i = 1, \ldots, n$, $e_i(t)$ also is CUUB. It further follows from Lemma lema2.2 that, for $k = 1, \ldots, N$ and $m = 1, \ldots, n$, the tracking error $\delta_{k,m}(t)$ is CUUB. Furthermore, Assumption 2.2 ensures the state boundedness of the leader. Hence, one has, for $k = 1, \ldots, N$, the follower's states $\bar{z}_k(t)$ are bounded, too. The proof is completed here.

2.4 Simulation

In this section, let us consider a class of leader-following systems, which contain one leader and five followers. The whole systems can be described by a digraph G, and the topology can be found in Fig. 2.1, where each number on each edge denotes the weight of the edge. The followers have the same forms in (2.2) which are described as follows:

$$\begin{cases} \dot{z}_{1,1}(t) = z_{1,2}(t) \\ \dot{z}_{1,2}(t) = z_{1,3}(t) \\ \dot{z}_{1,3}(t) = z_{1,2}\sin(z_{1,1}) + \cos{(z_{1,3})}^2 + u_1 + d_1 \end{cases}$$
$$\begin{cases} \dot{z}_{2,1}(t) = z_{2,2}(t) \\ \dot{z}_{2,2}(t) = z_{2,3}(t) \\ \dot{z}_{2,3}(t) = -z_{2,1}z_{2,2} + 0.01z_{2,1} - 0.01(z_{2,1})^2 + u_2 + d_2 \end{cases}$$
$$\begin{cases} \dot{z}_{3,1}(t) = z_{3,2}(t) \\ \dot{z}_{3,2}(t) = z_{3,3}(t) \\ \dot{z}_{3,3}(t) = z_{3,2} + \sin(z_{3,3}) + u_3 + d_3 \end{cases}$$
$$\begin{cases} \dot{z}_{4,1}(t) = z_{4,2}(t) \\ \dot{z}_{4,2}(t) = z_{4,3}(t) \\ \dot{z}_{4,3}(t) = -3(z_{4,1} + z_{4,2} - 1)^2(z_{4,1} + z_{4,2} + z_{4,3} - 1) \\ \qquad\qquad - z_{4,2} - z_{4,3} + 0.5\sin(2t) + \cos(2t) + u_4 + d_4 \end{cases}$$
$$\begin{cases} \dot{z}_{5,1}(t) = z_{5,2}(t) \\ \dot{z}_{5,2}(t) = z_{5,3}(t) \\ \dot{z}_{5,3}(t) = \cos(z_{5,1}) + u_5 + d_5 \end{cases}$$

and the leader is described in the form

Fig. 2.1 Topology of $\bar{G}$

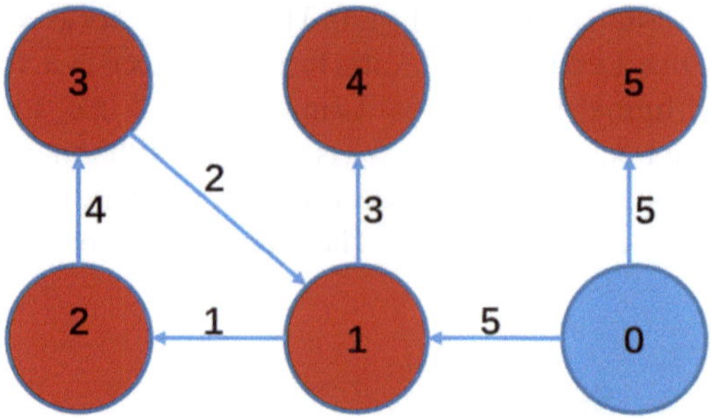

$$\begin{cases} \dot{z}_{0,1}(t) = z_{0,2}(t) \\ \dot{z}_{0,2}(t) = z_{0,3}(t) \\ \dot{z}_{0,3}(t) = -\frac{1}{3}(z_{0,1} + 4z_{0,2} + 3z_{0,3} - 1)(z_{0,1} + z_{0,2} - 1)^2 \\ \qquad - z_{0,2} - 2z_{0,3} + 6\cos(2t) + 3\sin(2t). \end{cases}$$

In this work, the random disturbance $|d_k| \leq 1$, for $k = 1, \ldots, N$.

From Fig. 2.1, it follows that $B = \begin{bmatrix} 5, & 0, & 0, & 0, & 0 \\ 0, & 0, & 0, & 0, & 0 \\ 0, & 0, & 0, & 0, & 0 \\ 0, & 0, & 0, & 0, & 0 \\ 0, & 0, & 0, & 0, & 5 \end{bmatrix}$. In this simulation, $\lambda_1 = 100$, $\lambda_1 = 20$, $\lambda_3 = 1$. Note that, it is assumed that, at $t = 20s$, in the 1st and 5th followers, their actuators are faulty and the actuators in the other followers are healthy. The actuator faults are $u_1^f = u_1 + 0.5\sin(t)$ and $u_5^f = u_5 + 0.5\sin(t)$.

Some simulation results are presented in Figs. 2.2, 2.3, 2.4, 2.5, 2.6, 2.7, 2.8, 2.9, 2.10, and 2.11. Figure 2.2 shows the state boundedness of the leader, which means that Assumption 2.1 is satisfied. Under normal operating condition (2.18), as shown in Figs. 2.3, 2.4, 2.5, 2.6, and 2.7, all the states of the followers are bounded. At the same time, Fig. 2.8 shows, each tracking error asymptotically converges to a neighborhood of zero. It is obvious from Figs. 2.2, 2.3, 2.4, 2.5, 2.6, 2.7, and 2.8 that the two results in this work have been verified. However, at time $t = 20s$, the faults occur in followers 1 and 5, if the normal controllers (2.18) still are adopted, and the tracking errors deviate significantly from the neighborhood of the origin, as shown in Fig. 2.9.

However, as long as the faults can be detected and compensated by the FTC defined in (2.29), good control performance is re-gained, which is shown in Fig. 2.10.

For the followers' control inputs, Fig. 2.11 shows their boundedness. From the above-mentioned simulation results, it is easy to find that the FTC method proposed in this work is effective.

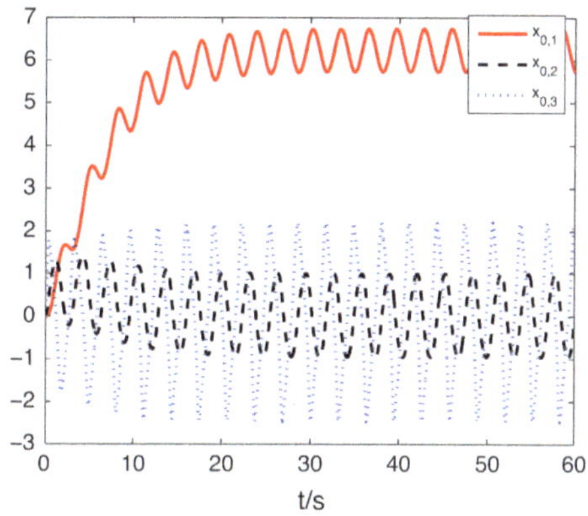

Fig. 2.2 States of the leader labeled 0

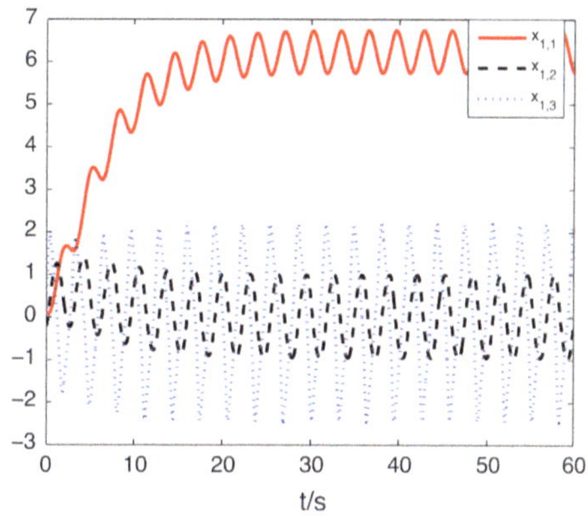

Fig. 2.3 States of follower 1

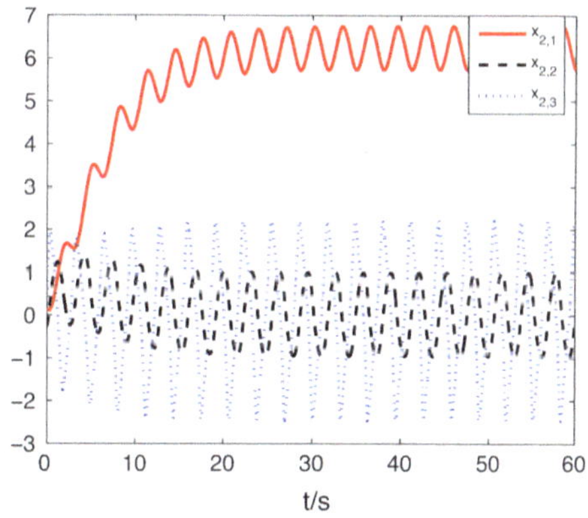

Fig. 2.4 States of follower 2

Fig. 2.5 States of follower 3

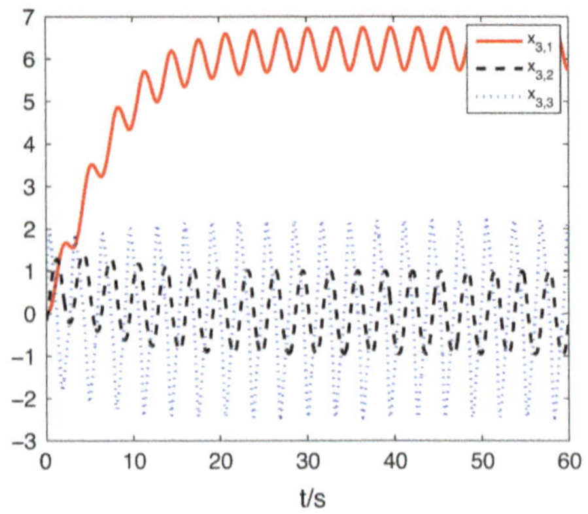

Fig. 2.6 States of follower 4

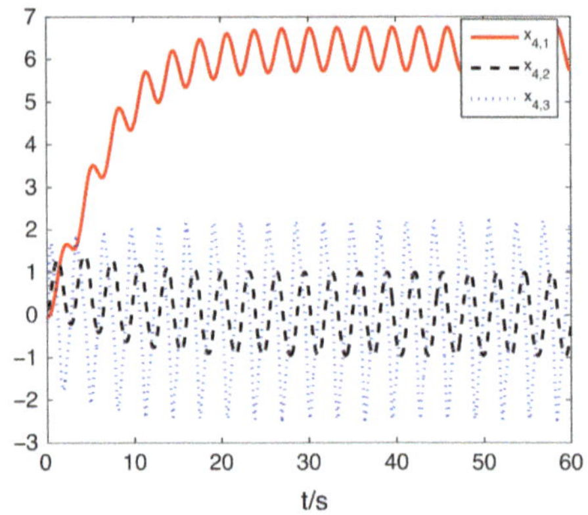

Fig. 2.7 States of follower 5

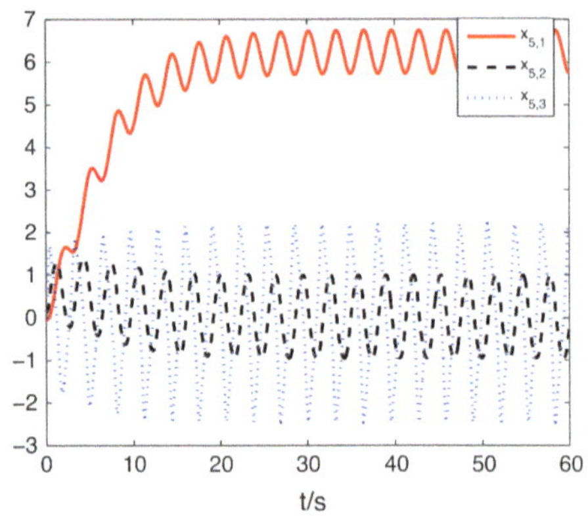

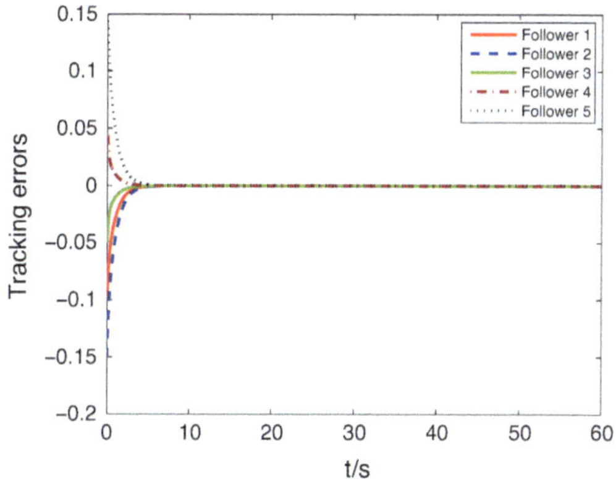

Fig. 2.8 Tracking errors under the fault-free condition

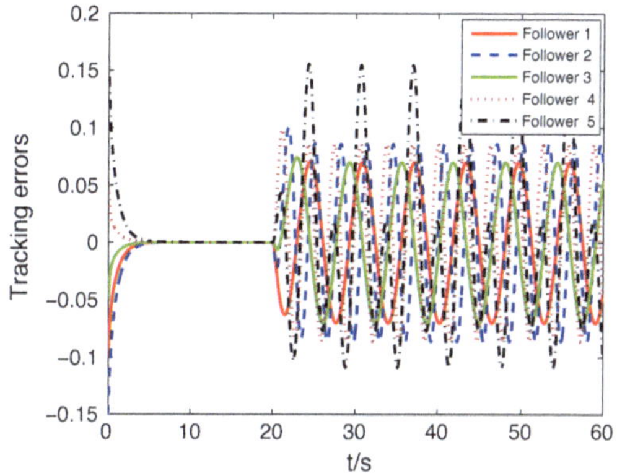

Fig. 2.9 Tracking errors under the faulty condition without FTC

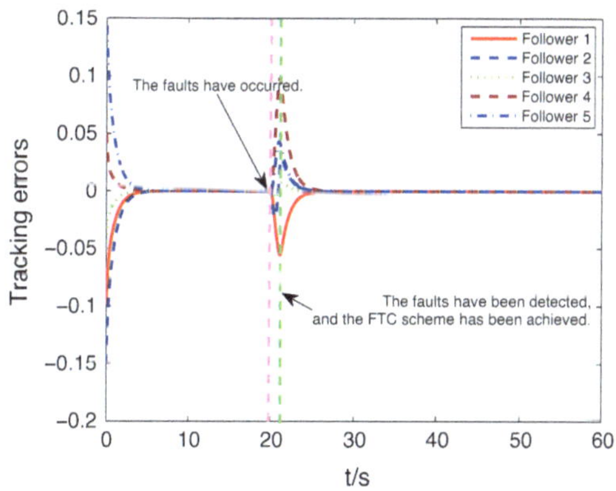

Fig. 2.10 Tracking errors with FTC

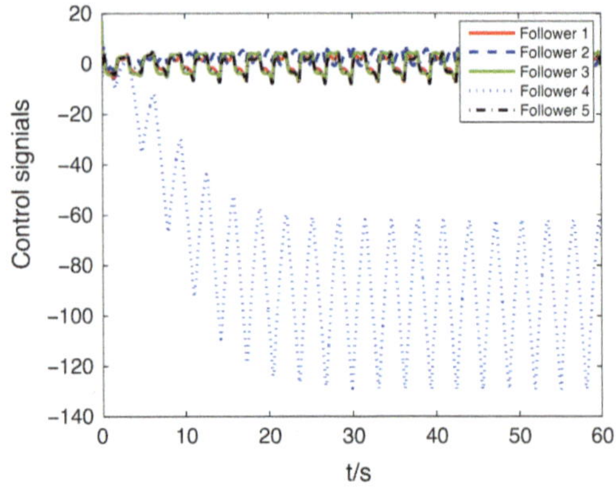

Fig. 2.11 Control signals

2.5 Conclusion

This chapter investigates the cooperative control problem of leader-following systems with actuator faults, and a novel distributed adaptive control scheme is proposed such that all followers can synchronize asymptotically to the leader and tracking errors between the leader and followers can converge to a small neighborhood of the origin in finite time. Furthermore, for the convergence time, its analytical expression is strictly derived and presented in this chapter. The stability analysis of the resulting closed-loop system is developed in sense of Lyapunov theory with algebraic graph theory. The FTC method proposed in this chapter not only reduces the influence of the faults impacted on system performance but also minimizes the delay raised by FD. Furthermore, the FTC method does not need the so-called FI module in the traditional active FTC method.

References

Defoort M, Floquet T, Kokosy A, Perruquetti W (2008) Sliding-mode formation control for cooperative autonomous mobile robots. IEEE Trans Ind Electron 55(11):3944–3953

Cruz D, McClintock J, Perteet B, Orqueda O, Cao Y, Fierro R (2007) Decentralized cooperative control-a multivehicle platform for research in networked embedded systems. IEEE Control Syst Mag 27(3):58–78

Wang X, Yadav V, Balakrishnan S (2007) Cooperative UAV formation flying with obstacle/collision avoidance. IEEE Trans Control Syst Technol 15(4):672–679

Chen J, Cao X, Cheng P, Xiao Y, Sun Y (2010) Distributed collaborative control for industrial automation with wireless sensor and actuator networks. IEEE Trans Ind Electron 57(12):4219–4230

Murray R (2007) Recent research in cooperative control of multivehicle systems. ASME J Dyn Syst Meas Control 129(5):571–583

Ren W, Cao C (2011) Distributed coordination of multi-agent networks?: emergent problems, models, and issues. Springer-Verlag, London

Zhang H, Lewis FL (2012) Adaptive cooperative tracking control of higher-order nonlinear systems with unknown dynamics. Autom 48(7):1432–1439

Ren W, Beard R, Atkins E (2007) Information consensus in multivehicle cooperative control. IEEE Control Syst Mag 27(2):71–82

Gao C, Wang Z, He X, Dong H (2022) Fault-tolerant consensus control for multiagent systems: an encryption-decryption scheme. IEEE Trans Autom Control 67(5):2560–2567

Zhu W, Cheng D (2010) Leader-following consensus of second-order agents with multiple time-varying delays. Autom 46(12):1994–1999

Wang W, Slotine J (2006) A theoretical study of different leader roles in networks. IEEE Trans Autom Control 51(7):1156–1161

Ren W, Moore K, Chen Y (2007) High-order and model reference consensus algorithms in cooperative control of multivehicle systems. ASME J Dyn Syst Meas Control 129(5):678–688

Hong Y, Chen G, Bushnell L (2008) Distributed observers design for leader-following control of multi-agent networks. Autom 44(3):846–850

Wang X, Chen G (2002) Pinning control of scale-free dynamical networks. Phys A 310(3–4):521–531

Khalil HK (2002) Nonlinear systems (3rd.). Upper Saddle River, NJ, Prentice Hall

Jiang F, Wang L, Jia Y (2009) Consensus in leaderless networks of high-order-integrator agents. In: Proceedings of the American Control Conference, Hyatt Regency Riverfront, St. Louis, MO, USA, June 10–12, pp 4458–4463

Ren W, Moore KL, Chen Y (2007) High-order and model reference consensus algorithms in cooperative control of multivehicle systems. ASME J Dyn Syst Meas Control 129(5):678–688

Wang J, Cheng D (2007) Consensus of multi-agent systems with higher order dynamics. In: Proceedings of the chinese control conference, July 26–31, Zhangjiajie, Hunan, China, pp 761–765

Zhang H, Lewis FL (2010) Synchronization of networked higher-order nonlinear systems with unknown dynamics. In: Proceedings of the 49th IEEE conference on decision and control, Dec. 15–17, Hilton Atlanta Hotel, Atlanta, GA, USA, pp 7129–7134

Zhang H, Lewis FL, Das A (2011) Optimal design for synchronization of cooperative systems: state feedback, observer and output feedback. IEEE Trans Autom Control 56(8):1948–1952

Zhang H, Lewis FL, Qu Z (2012) Lyapunov, adaptive, and optimal design techniques for cooperative systems on directed communication graphs. IEEE Trans Ind Electron 59(7):3026–3041

Yu H, Xia X (2012) Adaptive consensus of multi-agents in networks with jointly connected topologies. Autom 48:1783–1790

Bokovi JD, Li SM, Mehra RK (2002) Formation flight control design in the presence of unknown leader commands. In: Proceedings of the american control conference, Anchorage, AK, May 8–10, pp 2854–2859

Semsar-Kazerooni E, Khorasani K (2006) Adaptive formation control of UAVs in the presence of unknown vortex forces and leader commands. In: Proceedings of the american control conference, Minneapolis, Minnesota, USA, June 14–16, pp 3563–3568

Semsar-Kazerooni E, Khorasani K (2007) Optimal performance of a modified leader-follower team of agents with partial availability of leader command and presence of team faults. In: Proceedings of the 46th IEEE conference on decision and control, New Orleans, LA, USA, Dec. 12–14, pp 2491–2497

Semsar-Kazerooni E, Khorasani K (2008) Optimal consensus algorithms for cooperative team of agents subject to partial information. Autom 44(11):2766–2777

Bokovi JD, Bergstrom SE, Mehra RK (2005) Retrofit reconfigurable flight control in the presence of control effector damage. In: Proceedings of the American control conference, June 8–10, Portland, OR, USA, pp 2652–2657

Semsar-Kazerooni E, Khorasani K (2009) Analysis of actuator faults in a cooperative team consensus of unmanned systems. In: Proceedings of the American control conference, June 10–12, Hyatt Regency Riverfront, St. Louis, MO, USA, pp 2618–2623

Semsar-Kazerooni E, Khorasani K (2010) Team consensus for a network of unmanned vehicles in presence of actuator faults. IEEE Trans Control Syst Technol 18(5):1155–1161

Jiang B, Zhang K, Shi P (2010) Less conservative criteria for fault accommodation of time-varying delay systems using adaptive fault diagnosis observer. Int J Adapt Control Signal Process 24(4):322–334

Jiang B, Gao Z, Shi P, Xu Y (2010) Adaptive fault-tolerant tracking control of near-space vehicle using Takagi-Sugeno fuzzy models. IEEE Trans Fuzzy Syst 18(5):1000–1007

Wang LX (1994) Adaptive fuzzy systems and control: design and stability analysis. Prentice-Hall, Englewood Cliffs

Du H, Li S, Shi P (2012) Robust consensus algorithm for second-order multi-agent systems with external disturbances. Int J Control 85(12):1913–1928

Zhou Q, Shi P, Lu J, Xu S (2011) Adaptive output feedback fuzzy tracking control for a class of nonlinear systems. IEEE Trans Fuzzy Syst 19(5):972–982

Zhang J, Shi P, Xia Y (2010) Robust adaptive sliding mode control for fuzzy systems with mismatched uncertainties. IEEE Trans Fuzzy Syst 18(4):700–711

Tong SC, Li Y (2009) Observer-based fuzzy adaptive control for strict-feedback nonlinear systems. Fuzzy Sets Syst 160(12):1749–1764

Zhang TP, Wen H, Zhu Q (2010) Adaptive fuzzy control of nonlinear systems in pure-feedback form based on input-to state stability. IEEE Trans Fuzzy Syst 18(1):80–93

Zhang H, Shi Y, Saadat Mehr A (2012) On H_$infty$ filtering for discrete-time Takagi-Sugeno fuzzy systems. IEEE Trans Fuzzy Syst 20(2):396–401

Wu J, Shi Y (2011) Consensus in multi-agent systems with random delays governed by a Markov chain. Syst & Control Lett 60(11):863–870

Huang C (2012) Model based fuzzy control with affine T-S delayed models applied to nonlinear systems. Int J Innov Comput Inf Control 8(5(A)):2979–2993

Yordanova S, Jain LC (2012) Design of process fuzzy control for programmable logic controller. Int J Innov Comput Inf Control 8(12):8033–8048

Cai N, Xi J, Zhong Y, Ma H (2012) Controllability improvement for linear time-invariant dynamical multi-agent systems. Int J Innov Comput Inf Control 8(5(A)):3315–3328

Ejaz N, manzoor U, Nefti S, Baik S (2012) A collaborative multi-agent framework for abnormal activity detection in crowded areas. Int J Innov Comput Inf Control 8(6):4219–4234

Shen Q, Jiang B, Shi P (2013) Fault diagnosis for T-S fuzzy systems with sensor faults and system performance analysis. IEEE Trans Fuzzy Syst 21(2):289–300

Ghadiri H, Emami M, Khodadadi H (2021) Adaptive super-twisting non-singular terminal sliding mode control for tracking of quadrotor with bounded disturbances. Aerosp Sci Technol (112). https://doi.org/10.1016/j.ast.2021.106616

Xing W, Shi P, Song H, Zhao Y, Li L (2019) Global pinning synchronization of stochastic delayed complex networks. Inf Sci 490:113–125

Ni J, Shi P (2021) Adaptive neural network fixed-time leader-follower consensus for multi-agent systems with constraints and disturbances. IEEE Trans Cybern 51(4):1835–1848

Wang Q, Dong X, Yu J, Lü J, Ren Z (2021) Predefined finite-time output containment of nonlinear multi-agent systems with leaders of unknown inputs. IEEE Trans Circuits Syst I Regul Pap 68(8):3436–3448

Chapter 3
Distributed Cooperative Fuzzy Fault-Tolerant Backstepping Control Against Dead-Zone and Saturation

3.1 Introduction

In practical engineering, saturation and dead-zone are common phenomenon. These input nonlinearities are regarded by some experts and scholars as special actuator faults. Obviously, they will negatively impact the performance of the controlled systems. Hence, the input nonlinearities should be taken into account in control designing. For the input nonlinearities, most of the existing works reported in literature focus only on single-agent systems (Tao and Kokotovic 1994, 1995; Cho and Bai 1998; Zhang et al. 2016; Wang et al. 2004; Yau and Yan 2004; Hsu et al. 2004; Yan et al. 2005; Shyu et al. 2005) and does not concern multi-agent systems (MASs) considered in this chapter. In fact, to my best knowledge, up to now, the consensus control problem has not been studied fully yet for MASs subject to the input nonlinearities. However, the research about MASs with the nonlinearities is very practically useful and very important in both applications and theory, which thus motivates us for the work in this chapter.

In past 10 years, for single-agent systems, saturation, dead-zone, and the other input nonlinearities have been extensively studied by various scholars, and a lot of control design strategies are designed to reduce or cancel the nonlinearities' influence on control performance (Tao and Kokotovic 1994, 1995; Cho and Bai 1998; Zhang et al. 2016; Wang et al. 2004; Yau and Yan 2004; Hsu et al. 2004; Yan et al. 2005; Shyu et al. 2005; Ahn 2013, 2014, 2015a, b).

Notice that, in most of the existing results reported in literature, a basic assumption is made, namely, only one input nonlinearity occurs in a system. For example, the results such as (Tao and Kokotovic 1994, 1995; Yan et al. 2005; Shyu et al. 2005) investigated only dead-zone, and the other input nonlinearities such as sector input and saturation were not involved. In practice, however, multiple input nonlinearities may occur simultaneously in a system. In addition, in the existing works, dead-zone model is often assumed to be symmetrical. In real applications, the assumption does not always hold. Thus, multiple different input nonlinearities must be considered

Q. Shen, *Distributed Fault-Tolerant Consensus Control of Leader-Following Systems*, Intelligent Control and Learning Systems 11,
https://doi.org/10.1007/978-981-99-7426-9_3

simultaneously and compensated in designing control schemes for dynamic systems, especially nonlinear uncertain leader-following systems. How to compensate for the nonlinearities is challenging and still open.

For MASs including the leader-following systems, although there are many different research directions, their control problems always are one hot issues. For the last two decades, the problems have attracted extensive attention from different areas (Chen et al. 2010; Wang and Yu 2016; Ren and Cao 2011; Qu 2009; Olfati-Saber et al. 2007), and fruitful results have been obtained. For the consensus of the leader-following systems, distributed consensus control design have inspired great interest. In the studies of the consensus, the consensus can generally be divided into two types: output consensus and state (Shen et al. 2014; Shen and Shi 2015, 2016a, b; Shi and Shen 2015; Shen et al. 2020; Yu and Xia 2012; Ren et al. 2007; Zhang and Lewis 2012; Fiorelli et al. 2006; Zhu and Cheng 2010). At the first stage of the research, simple systems were investigated. In recent years, researchers began to focus on high-order MASs including certain and uncertain systems, and lots of results have been reported in literature (Shen et al. 2014; Shen and Shi 2015, 2016a, b; Shi and Shen 2015). Most of the results, however, were obtained under the assumption that the actuators in the controlled systems were not subject to the input nonlinearities. However, as mentioned earlier, input nonlinearity is a common phenomenon that is widely found in real engineering applications. What's worst, the nonlinearity can reduce control accuracy and further affect control performance. Therefore, for the leader-following consensus, the actuator nonlinearities should be considered and compensated.

In this chapter, for a class of nonlinear uncertain leader-following systems, with both saturation and dead-zone considered, the consensus is investigated. By combining FLSs and backstepping design with adaptive control, an distributed control method is proposed such that the consensus can be achieved.

In this work, main contribution deserves to be highlighted.

(i) For a class of LFSs, a more realistic scenario is studied in this chapter, multiple input nonlinearities or faults in actuators are considered.
(ii) Multiple nonlinearities or faults in actuators, i.e., dead-zone and saturation, are investigated simultaneously in this chapter. Furthermore, model parameters in the nonlinearities and dead-zone are not assumed to be known and symmetric, respectively.
(iii) In addition, for the disturbance and approximation errors, it is unnecessary that their upper boundaries are known in prior.

3.2 Preliminaries and Problem Formulation

Some preliminaries are introduced firstly in this section. Then, for a class of leader-following systems, their consensus problems are formulated. Finally the introduction of fuzzy logic systems (FLSs) is given.

3.2.1 Preliminaries

In the sense of mathematics, MASs including LFSs investigated in this book can be represented by a graph. Obviously, it involves basic graph theory in the research of MASs. So, let us first introduce it in this subsection. For basic graph theory, it was already introduced in the previous chapter. For the convenience of reading, its related knowledge will be re-introduced here again. If the readers have gained an understanding of the basic graph theory, the readers can skip this part and read the following content directly.

A class of multi-agent systems will be considered here, which consist of N followers and a leader. For the leader and followers, the information exchange among the leader and followers is often modeled as a graph. Hence, we first introduce the knowledge of basic graph theory in this subsection.

A directed graph can be represented as $\mathscr{G} = (\mathscr{V}, \mathscr{E})$, where agent set $\mathscr{V}$ and edge (arcs) set $\mathscr{E}$ are defined as $\mathscr{V} = (\mathscr{V}_1, \ldots, \mathscr{V}_N)$ and $\mathscr{E} \subseteq \mathscr{V} \times \mathscr{V}$, respectively. Note that, if there exists an edge from node i to node j, then $(\mathscr{V}_i, \mathscr{V}_j) \in E$. For $\mathscr{G}$, its adjacency matrix $\mathscr{A}$ can be described as $\mathscr{A} = [a_{ij}] \in R^{N\times N}$, where element a_{ij} is defined as follows: (1) $a_{ij} = 0$ if there exists no an edge from node i to node j, i.e., $(\mathscr{V}_i, \mathscr{V}_j) \notin \mathscr{E}$; and (2) $a_{ij} \neq 0$ if there exists no an edge from node i to node j, i.e., $(\mathscr{V}_i, \mathscr{V}_j) \in E$. Throughout this chapter, a_{ii} is assumed to be equal to zero, i.e.,$a_{ii} = 0$, and the topology is also assumed to be fixed, which means that $\mathscr{A}$ is fixed and is not variant.

In directed graph $\mathscr{G}$, for the i-th agent, its in-degree d_i is defined as $d_i = \sum_{j=1}^{N} a_{ij}$. Correspondingly, for the directed graph $\mathscr{G}$, its in-degree matrix and Laplacian matrix, i.e., $\mathscr{D}$ and $\mathscr{L}$, are also defined, respectively, as $\mathscr{D} = diag\{d_1, \ldots, d_N\} \in R^{N\times N}$ and $\mathscr{L} = [l_{ij}] = \mathscr{D} - \mathscr{A} \in R^{N\times N}$. For the Laplacian matrix $\mathscr{L}$, $L\underline{1} = 0$ with $\underline{1} = [1, \ldots, 1]^T \in R^{N\times 1}$. In addition, for agent i, its neighbor set can be described as $N_i = \{j|(\mathscr{V}_j, \mathscr{V}_i) \in E\}$. Obviously, for the i-th agent, if the j-th agent is its neighbor, agent i can obtain the information from the j-th agent. For directed graph, however, not necessarily vice versa. For undirected graph, neighbors are interrelated. From node i to node j, if there exists a sequence of successive arcs, which can be described in the form $\{(\mathscr{V}_i, \mathscr{V}_l), (\mathscr{V}_l, \mathscr{V}_k), \ldots, (\mathscr{V}_m, \mathscr{V}_j)\}$, then a direct path exists. In a digraph, if there exist a special node and a directed path from the special node to every other node, then graph is said to have a spanning tree. For any two different nodes $\mathscr{V}_i$ and $\mathscr{V}_j$ in the digraph, from $\mathscr{V}_i$ to $\mathscr{V}_j$, if there always exists a direct path, then the graph is said to be strongly connected.

Let $\bar{\mathscr{E}} \subseteq \bar{\mathscr{V}} \times \bar{\mathscr{V}}$ and $\bar{\mathscr{V}} = \{\mathscr{V}_0, \mathscr{V}_1, \mathscr{V}_2, \ldots, \mathscr{V}_N\}$. Then, based on the graph $\mathscr{G}$ and a leader node, a new graph $\bar{\mathscr{G}}$ is defined as $\bar{\mathscr{G}} = \{\bar{\mathscr{V}}, \bar{\mathscr{E}}\}$. Note that, from the leader node to node k, if there exists an edge, then its weight $b_k > 0$, otherwise $b_k = 0$, where $k = 1, 2, \ldots, N$.

Now, the following lemma and assumption are made.

Lemma 3.1 (Zhang and Lewis 2012) *Define* $q = (\mathscr{B} + \mathscr{L})^{-1}\underline{1} = [q_1, \ldots, q_N]$,

$$P = diag\{\frac{1}{q_i}\} \in R^{N\times N} \;\; with \;\; p_i = \frac{1}{q_i}, \quad Q = P(\mathscr{B} + \mathscr{L}) + (\mathscr{B} + \mathscr{L})^T P,$$

then matrices Q *and* P *are positive definite, i.e.,* $Q > 0$ *and* $P > 0$.

Assumption 3.1 There exists a spanning tree in graph $\bar{\mathscr{G}}$, where the leader is the root, and its topology is invariance.

3.2.2 Problem Formulation

Consider the following class of nonlinear systems, which have N nodes and are modeled as graph $\mathscr{G}$. Each node in $\mathscr{G}$ is seen as a follower and is labeled k, $k = 1, \ldots, N$. The k-th follower has the following form:

$$\begin{cases} \dot{\pi}_k = \varpi_k \\ \dot{\varpi}_k = f_k(\varpi_k) + g_k u_k + h_k(\pi_k, \varpi_k) \\ y_k = \pi_k, \end{cases} \tag{3.1}$$

where $\pi_k \in R$ and ϖ_k are system states, $u_k \in R$ is system input, and y_k system output;

nonlinear function $f_k(\varpi_k)$ is unknown but smooth; control gain g_k is an unknown real number; and function $h_k(\cdot)$ denotes system uncertainties, modeling errors, and so on.

In addition, there exists a node, which is labeled 0 and seen as leader node. Its dynamics is

$$\begin{cases} \dot{\pi}_0 = \varpi_0 \\ \dot{\varpi}_0 = f_0(\varpi_0, t) \\ y_0 = \pi_0, \end{cases} \tag{3.2}$$

where $\pi_0 \in R$ and $\varpi_0 \in R$ are the states of the leader, nonlinear function $f_0(\varpi_0, t) \in R$ is unknown for each follower in $\mathscr{G}$.

Based on $\mathscr{G}$ and the leader, a new network $\bar{\mathscr{G}}$ is constructed, which considered the followers in $\mathscr{G}$ and the leader. The descriptions and definitions about graph $\bar{\mathscr{G}}$ can be found in the previous subsection.

In this chapter, the main task is to construct a control strategy for the leader-following systems modeled by graph $\bar{\mathscr{G}}$ to guarantee the output asymptotical synchronization between the leader and followers is achieved, namely, for $\forall m \in \{1, \ldots, N\}$, $\pi_m \to \pi_0$.

Although lots of works about synchronization or consensus control, for the leader-following systems, most of these works are carried out under ideal conditions that system actuators have no input linearity or fault. In this chapter, a more practical case is considered, where actuator faults or input nonlinearities occurred in the controlled systems. In practice, however, input nonlinearities, such as saturation, dead-zone, sector, and so on, are inevitably present in the actuator. Worst of all, these nonlinearities can degrade control performance and even lead to more serious consequences. Therefore, for the input nonlinearities mentioned above, their influence must be taken into account in stability analysis and control designing for dynamic systems, especially the leader-following systems studied in this book.

In this chapter, two linearities, i.e., saturation and dead-zone occur simultaneously in the followers. In the faulty case, the dynamics of the followers have the following form:

$$\begin{cases} \dot{\pi}_k = \varpi_k \\ \dot{\varpi}_k = f_k(\varpi_k) + g_k sat_k(dz_k(u_k)) + h_k, \end{cases} \tag{3.3}$$

where π_k, ϖ_k, and u_k, respectively, denote the states and input; $f_k(\varpi_k)$ and h_k are defined in the same way as (3.1); and saturation function$sat(\cdot) \in R$ and dead-zone function $dz_k(\cdot) \in R$ are introduced in the following subsection.

To design desired distributed consensus protocols for followers (3.3), the following assumption is introduced in this chapter.

Assumption 3.2 Function $|h_k| \leq M_{h_k}$ with constant $M_{h_k} > 0 \in R$ being unknown, $\forall k \in \{1, \ldots, N\}$.

3.2.3 Descriptions of Saturation and Dead-Zone

Let us first introduce dead-zone, which is shown in Fig. 3.1. It can be described as follows:

$$dz(u) = \begin{cases} c_-(u + u_-), & u < -u_- \\ 0, & -u_- \leq u \leq u_+ \\ c_+(u - u_+), & u > u_+, \end{cases} \tag{3.4}$$

where model constants $c_l > 0$ and $u_l > 0$, $l = +, -$. Saturation shown in Fig. 3.2 has the following form:

$$\begin{aligned} sat(u) &= \begin{cases} -u_L, & u < -u_L \\ u, & -u_L \leq u \leq u_H \\ u_H, & u > u_H \end{cases} \\ &= \omega(u)u, \end{aligned} \tag{3.5}$$

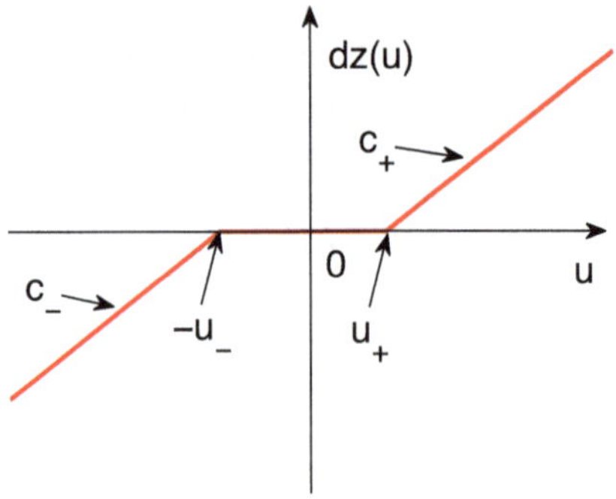

Fig. 3.1 Dead-zone function $dz(u)$

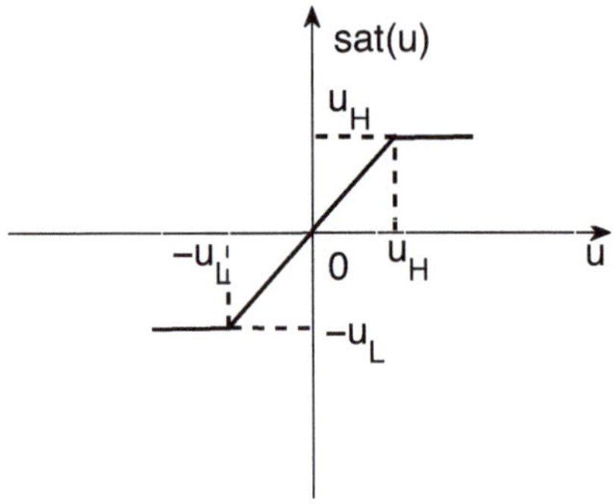

Fig. 3.2 Saturation function $sat(u)$

where constants $u_m > 0$, $m = H, L$, and $\omega(u)$ have the following form:

$$\omega(u) = \begin{cases} -\dfrac{u_L}{u}, & u < -u_L \\ 1, & -u_L \leq u \leq u_H \\ \dfrac{u_H}{u}, & u > u_H. \end{cases} \tag{3.6}$$

From (3.6), it is easy to find that, $\omega(u) \in R$ has the following property that $\tau < \omega(u) \leq 1$ with constant $\tau > 0$ being unknown.

Notice that, dead-zone and saturation occur simultaneously in each follower in graph $\bar{\mathscr{G}}$. The integrated linearity is shown in Fig. 3.3. In this work, an assumption that $c_-u_- < u_L$ and $c_+u_+ < u_H$. Then, from (3.4)–(3.6), the integrated linearity $sat(dz(u))$ is represented as

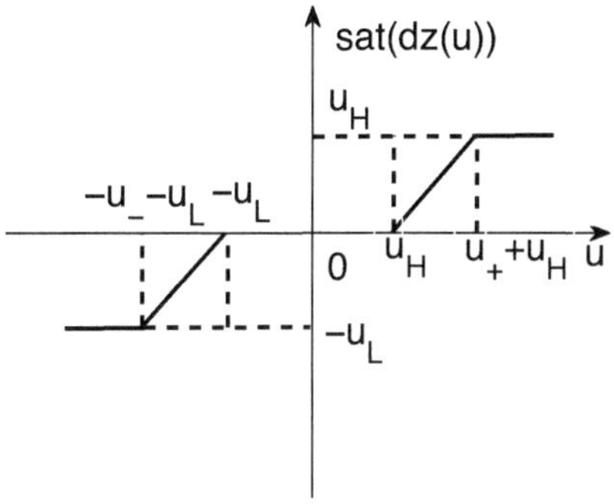

Fig. 3.3 Saturation dead-zone function $sat(dz(u))$

$$
sat(dz(u)) = \begin{cases} \dfrac{u_L}{dz(u)} dz(u) = \omega(dz(u)) c_-(u + u_-), & u < -\dfrac{u_L}{c_-} - u_- \\ c_-(u + u_-), & -\dfrac{u_L}{c_-} - u_- \leq u < -u_- \\ 0, & -u_- \leq u \leq u_+ \\ c_+(u - u_+), & \dfrac{u_H}{c_+} + u_+ \geq u > u_+ \\ \dfrac{u_H}{dz(u)} dz(u) = \omega(dz(u)) c_+(u - u_+), & u > \dfrac{u_H}{c_+} + u_+. \end{cases} \tag{3.7}
$$

Note that this work involves not one follower but multiple followers. Hence in order to distinguish these followers as well as their respective parameters, let us add subscript k to the symbols in (3.7). For example, for the k-th follower, the symbols in (3.7) have been re-described as follows: $sat_k(dz_k(u_k))$, $\omega_k(dz_k(u_k))$, $c_{-,k}$, $c_{+,k}$, $u_{L,k}$, $u_{H,k}$, $u_{-,k}$, $u_{+,k}$, and τ_k.

3.2.4 Fuzzy Logic Systems

In this subsection, let us first introduce briefly fuzzy logic systems (FLSs). Fuzzy logic systems, as an universal approximation, have been utilized in control designing.

For an FLS, in general, we have four necessary parts: (1) fuzzifier, (2) defuzzifier, (3) knowledge base, and (4) fuzzy inference engine. Note that, it is on fuzzy rules that the fuzzy inference engine is working. The fuzzy rules can be described as

$$R^l : \text{if } z_1 \text{ is } \Pi_1^l \text{ and } z_2 \text{ is } \Pi_2^l \cdots \text{ and } z_n \text{ is } \Pi_n^l \text{ then } y \text{ is } \Xi^l, l = 1, 2, \ldots, M,$$

where y and $\underline{z} = [z_1, z_2, \ldots, z_n]^T \in U \subset R^n$ denote, respectively, the output and input of the FLS. The symbols Π_i^l and Ξ^l denote fuzzy sets, fuzzy functions

$\mu_{\Pi_i^l}(z_i)$, and $\mu_{\Xi^l}(y^l)$ can be described, respectively, as $\mu_{\Pi_i^l}(z_i) = \exp(-(\frac{z_i - a_i^l}{b_i^l})^2)$ and $\mu_{\Xi^l}(y^l) = 1$, respectively. The rule number is denoted by M.

Note that, in an FLS, its knowledge base consists of a collection of the rules mentioned above.

Through the following procedures: singleton function, center average defuzzification, and product inference, the above-mentioned FLS is represented in the following form:

$$y(z) = \frac{\left\{\sum_{l=1}^{M} \bar{y}^l \left(\prod_{i=1}^{n} \mu_{\Pi_i^l}(z_i)\right)\right\}}{\left\{\sum_{l=1}^{M} \left(\prod_{i=1}^{n} \mu_{\Pi_i^l}(z_i)\right)\right\}}, \tag{3.8}$$

where $\bar{y}^l = \max_{y \in R} \mu_{\Xi^l}$. Now, the fuzzy basis functions are defined as follows:

$$S_l(z) = \frac{\left[\prod_{i=1}^{n} \mu_{\Pi_i^l}(z_i)\right]}{\left[\sum_{l=1}^{M} \left(\prod_{i=1}^{n} \mu_{\Pi_i^l}(z_i)\right)\right]} \tag{3.9}$$

and define $S(z) = [S_1(z), \ldots, S_M(z)]^T$ and $\theta^T = [\bar{y}^1, \bar{y}^2, \ldots, \bar{y}^M] = [\theta_1, \theta_2, \ldots, \theta_M]$, then the above FLS is represented as

$$y(z) = \theta^T S(z).$$

Lemma 3.2 (Ahn 2015a) *On a compact set Ω, a function $f(z)$ is continuous. Then for any constant $\varepsilon > 0$, there exists a fuzzy logic system satisfying the condition*

$$\sup_{z \in \Omega} |f(z) - \theta^T S(z)| \le \varepsilon.$$

By Lemma 3.2, on a compact set, a continuous nonlinear function $f(z)$ is approximated in the following form:

$$f(z, \theta) = \theta^T S(z).$$

Optimal parameter vector θ^* is defined as follows:

$$\theta^* = \arg\min_{\theta \in \Omega}[\sup_{z \in U} |f(x) - \theta^T S(z)|],$$

where compact sets U and Ω are associated, respectively, with states z and θ. For the function $f(z)$, its minimum approximation error is

$$\varepsilon = f(z) - \theta^{*T} S(z). \tag{3.10}$$

In this work, fuzzy logic system will be utilized to approximate $f_k'(\bar{x}_k)$ by the following form:

$$f_k'(\bar{x}_k) = \theta_k^{*T} S_k(\bar{x}_k) + \varepsilon_k(\bar{x}_k),$$

where $\bar{x}_k = [x_{k,1}, \ldots, x_{k,m_k}]^T = [\pi_k, \varpi_k, \pi_j, \varpi_j, 1]^T$, $j \in N_k$.

Note that the unknown smooth function $f_k'(\bar{x}_k)$ is derived in the following analysis, which will be defined later.

Assumption 3.3 Both of sector θ_k^* and minimum approximation error $\varepsilon_k(\bar{x}_k)$ are bounded over a compact set, and $|\varepsilon_k(\bar{x}_k)| \leq M_{\varepsilon k}$ with unknown constant $M_{\varepsilon k} > 0$, for $k = 1, \ldots, N$.

3.3 Main Results

To achieve the consensus between the leader and followers, backstepping technology is used to design control scheme for the leader-following system (3.1) and (3.3) with dead-zone and saturation.

Here, for follower k ($k = 1, 2, \ldots, N$), let us define its neighborhood synchronization error e_k,

$$e_k = \sum\nolimits_{j \in N_k} a_{kj}(\pi_j - \pi_k) + b_k(\pi_0 - \pi_k). \tag{3.11}$$

For the leader-following system defined by (3.1) and (3.2), its global synchronization error e is defined as

$$e = [e_1, \ldots, e_N]^T. \tag{3.12}$$

Further,

$$\dot{e} = -(\mathscr{L} + \mathscr{B})(\pi - \underline{\pi}_0), \tag{3.13}$$

where $\pi = [\pi_1, \ldots, \pi_N]^T$ and $\underline{\pi}_0 = [\pi_0, \ldots, \pi_0]^T \in R^N$.

Then, for follower k ($k = 1, 2, \ldots, N$), let us define its tracking error as

$$\delta_k = \pi_k - \pi_0.$$

Let $\delta = [\pi_1, \ldots, \pi_N]^T \in R^N$, then

$$\delta = \pi - \underline{\pi}_0.$$

For the tracking error vector δ, a lemma is introduced here, which will be used in the following theoretical analysis.

Lemma 3.3 (Zhang and Lewis 2012) *For the tracking error vector δ and global synchronization error vector e, the condition holds,*

$$||\delta|| \leq ||e||/\underline{\sigma}(\mathscr{L} + \mathscr{B}),$$

where symbol $\underline{\sigma}(\bullet)$ denotes the minimum singular value of matrix $\bullet$.

It follows from (3.2), (3.3), and (3.11) that

$$\begin{aligned} \dot{e}_k &= \sum\nolimits_{j \in N_k} a_{kj}(\varpi_j - \varpi_k) + b_k(\varpi_0 - \varpi_k) \\ &= \sum\nolimits_{j \in N_k} a_{kj}\varpi_j + b_k\varpi_0 - (d_k + b_k)\varpi_k. \end{aligned} \tag{3.14}$$

Further,

$$\begin{cases} \dot{e}_k = \sum\nolimits_{j \in N_k} a_{kj}\varpi_j + b_k\varpi_0 - (d_k + b_k)\varpi_k \\ \dot{\varpi}_k = f_k(\varpi_k) + g_k u_k + h_k. \end{cases} \tag{3.15}$$

In the following control designing, backstepping control technique will be used to achieve the consensus between the leader and followers.

For $k = 1, \ldots, N$, define

$$r_{k,1} = e_k, \; r_{k,2} = \varpi_k - \alpha_k, \tag{3.16}$$

where α_k denotes a virtual control, which is a smooth function and is defined in Step 1.

Note that, in the following analysis, a function $f_k'(\bar{x}_k)$ will be derived, which is defined as $f_k'(\bar{x}_k) = f_k(\varpi_k) - \dot{\alpha}_k$. Further, from the later analysis, it can be obtained that the function $f_k'(\bar{x}_k)$ is unknown but continuous. Hence, it can be approximated by an FLS as follows:

$$f_k'(\bar{x}_k) = \theta_k^{*T} S_k(\bar{x}_k) + \varepsilon_k(\bar{x}_k),$$

where $\bar{x}_k = [dz_k, \varpi_k, dz_j, r_j, 1]^T$, $j \in N_k$.

For follower k, define control and adaptive laws as

$$u_k = \begin{cases} -u_k'(1 + \hat{\tau}_{c,k}) - u_{-,k}, & r_{k,2} \geq 0 \\ u_k'(1 + \hat{\tau}_{c,k}) + u_{+,k}, & r_{k,2} < 0; \end{cases} \tag{3.17}$$

$$u_k' = |\hat{\theta}_k S_k| + (1 + \frac{d_k + b_k}{2})|r_{k,2}| + \hat{M}_{h\varepsilon k} \tag{3.18}$$

$$\dot{\hat{\tau}}_{c,k} = \xi_\tau \hat{\tau}_{c,k}^3 u_k' |r_{k,2}|, \quad \hat{\tau}_{c,k}(0) > 0 \tag{3.19}$$

$$\dot{\hat{M}}_{h\varepsilon k} = \xi_M |r_{k,2}|, \quad \hat{M}_{h\varepsilon k}(0) > 0 \tag{3.20}$$

$$\dot{\hat{\theta}}_k = \xi_\theta r_{k,2} S_k, \quad \hat{\theta}_k(0) > 0, \tag{3.21}$$

where $c_k = \min\{c_{+,k}, c_{-,k}\}$, $\hat{\tau}_{c,k} > 0$ is the estimate of $\tau_{c,k}$, $\tau_{c,k} = g_k \tau_k c_k$, design parameters $\xi_\theta > 0$, $\xi_M > 0$ and $\xi_\tau > 0$, $u'_k \geq 0$. $\hat{M}_{h\varepsilon k}$ and $\hat{M}_{h\varepsilon k}$ denote, respectively, the estimates of unknown constants $M_{h\varepsilon k}$ and $M_{h\varepsilon k}$. Note that, u'_k, $M_{h\varepsilon k}$, and $M_{h\varepsilon k}$ will be defined later.

Now the main results in this work are given by the following theorem.

Theorem 3.1 *Consider the leader-following system defined by (3.2) and (3.3) with Assumptions 3.1–3.3. Under the condition that the followers are subject to saturation and dead-zone described by (3.4), (3.5), and (3.6), by applying the control input described by (3.17) and (3.18) with the adaptive laws described by (3.19)–(3.21). If the parameter $l_k > 0$ is chosen appropriately to guarantee the following inequality holds:*

$$l_k \geq \frac{1}{2}(d_k + b_k), \tag{3.22}$$

then the output synchronization between the leader and followers can be obtained and each synchronization error converges asymptotically to the origin.

Proof **Step 1**: Let us define $\bar{r}_1 = (r_{1,1}, \ldots, r_{N,1})^T$.

Choose matrix $P = diag\{p_1, \ldots, p_N > 0\}$ in Lemma 3.1, and define the following Lyapunov function:

$$V_1 = \frac{1}{2}\bar{r}_1^T P \bar{r}_1 = \sum\nolimits_{k=1}^{N} \frac{1}{2} p_k r_{k,1}^2.$$

From (3.15) and (3.16), the time derivative of V_1 can be developed

$$\begin{aligned} \dot{V}_1 &= \sum\nolimits_{k=1}^{N} r_{k,1}[\sum\nolimits_{j\in N_k} a_{kj}\varpi_j - (b_k + d_k)\varpi_k + b_k r_0] \\ &= \sum\nolimits_{k=1}^{N} \{r_{k,1}[\sum\nolimits_{j\in N_k} a_{kj}\varpi_j + b_k r_0] - \\ &\quad (b_k + d_k) r_{k,1}\alpha_\iota - (b_k + d_k) r_{k,1} r_{k,2}\}. \end{aligned} \tag{3.23}$$

Further, it follows from Young's inequality that

$$\begin{aligned} & - (b_k + d_k) r_{k,1} r_{k,2} \\ & \leq \frac{1}{2}(b_k + d_k) r_{k,1}^2 + \frac{1}{2}(b_k + d_k) r_{k,2}^2. \end{aligned} \tag{3.24}$$

From (3.23) and (3.24), one has

$$\dot{V}_1 \leq \sum_{k=1}^{N} \{r_{k,1}[\sum_{j\in N_k} a_{kj}\varpi_j + b_i\varpi_0]- \\ (d_k + b_k)r_{k,1}\alpha_i + \frac{1}{2}(d_k + b_k)r_{k,1}^2 + \\ \frac{1}{2}(d_k + b_k)r_{k,2}^2\}. \tag{3.25}$$

Here, the following virtual control, i.e., α_k, is designed for the dynamics of r_k as follows:

$$\alpha_k = \frac{1}{(d_k + b_k)}(l_k r_{k,1} + \sum_{j\in N_k} a_{kj}\varpi_j + b_k\varpi_0), \tag{3.26}$$

where design parameter $l_k > 0$ is chosen appropriately to guarantee the following inequality holds:

$$l_k \geq \frac{1}{2}(d_k + b_k).$$

Substituting (3.26) into (3.25), we have

$$\dot{V}_1 \leq \sum_{k=1}^{N} \{-(l_k - \frac{b_k + d_k}{2})r_{k,1}^2 + \frac{(b_k + d_k)}{2} r_{k,2}^2\}. \tag{3.27}$$

Step 2: Now, the following equation is considered:

$$\dot{\varpi}_k = f_k(\varpi_k) + h_k + g_k u_k.$$

Define

$$V_2 = V_1 + \sum_{k=1}^{N} \frac{1}{2} r_{k,2}^2.$$

From (3.15) and (3.27), its time derivative of V_2 can be described as

$$\dot{V}_2 = \dot{V}_1 + \sum_{k=1}^{N} r_{k,2}(h_k + g_k sat_k(dz_k(u_k)) + f_k - \dot{\alpha}_k). \tag{3.28}$$

Recall control law for follower k in (3.17),

$$u_k = \begin{cases} -u_k'(\hat{\tau}_{c,k} + 1) - u_{-,k}, & r_{k,2} \geq 0 \\ u_k'(\hat{\tau}_{c,k} + 1) + u_{+,k}, & r_{k,2} < 0, \end{cases}$$

where $c_k = \min\{c_{-,k}, c_{+,k}\}$, $\tau_{c,k} = g_k \tau_k c_k$, $\hat{\tau}_{c,k} > 0$ is the estimate of $\tau_{c,k}$, and $u_k' \geq 0$ defined later.

Note that, from two scenarios, the following analysis is represented.

Case 1:$r_{k,2} < 0$

Due to $r_{k,2} < 0$, from (3.17), it is further obtained

$$u_k = u_k'(1 + \hat{\tau}_{c,k}) + u_{+,k}.$$

For $\hat{\tau}_{c,k}$, from the definition of its adaptive law defined in (3.19), we know, $\hat{\tau}_{c,k} > 0$, for $t > 0$. And since $u_{+,k} > 0$, we further have $u_k > u_{+,k} > 0$. Thus, it follows from (3.4) that $dz_k(u_k) = c_{+,k}(u_k - u_{+,k})$.

From (3.7), it follows **(I)** if $u_k < \frac{u_{H,k}}{c_{+,k}} + u_{+,k}$, then $dz_k(u_k) = c_{+,k}(u_k - u_{+,k})$. Obviously, $dz_k(u_k) < u_{H,k}$. So, it follows from (3.7) that

$$\begin{aligned} sat_k(dz_k(u_k)) &= dz_k(u_k) \\ &= c_{+,k}(u_k - u_{+,k}). \end{aligned}$$

Therefore, substituting the control law (3.17) and (3.18) into the term $g_k r_{k,2}$ $sat_k(dz_k(u_k))$ in (3.28), we further have

$$\begin{aligned} g_k r_{k,2} sat_k(dz_k(u_k)) &= c_{+,k} g_k r_{k,2}(u_k - u_{+,k}) \\ &= c_{+,k} u_k' g_k r_{k,2}(1 + \hat{\tau}_{c,k}) \\ &= c_{+,k} g_k r_{k,2} u_k' + c_{+,k} g_k r_{k,2} u_k' \hat{\tau}_{c,k}. \end{aligned} \tag{3.29}$$

In this work, parameter $c_{+,k}$ is assumed to be positive, i.e., $c_{+,k} > 0$. In addition, in this case, $r_{k,2}$ is negative, namely, $r_{k,2} < 0$. From the control law u_k in (3.17), we have $\hat{\tau}_{c,k} > 0$ and $u_k' \geq 0$. So, we have

$$r_{k,2} = -|r_{k,2}|$$

$$c_{+,k} u_k' \hat{\tau}_{c,k} > 0, \;\; c_{+,k} u_k' > 0.$$

From (3.7), we known $1 > \omega_k(dz_k(u_k)) > \tau_k > 0$. Hence, (3.29) can further developed as

$$\begin{aligned} & g_k r_{k,2} sat_k(dz_k(u_k)) \\ & = -c_{+,k} g_k |r_{k,2}| u_k' - c_{+,k} g_k |r_{k,2}| u_k' \hat{\tau}_{c,k} \\ & \leq -c_{+,k} g_k |r_{k,2}| u_k' \hat{\tau}_{c,k} \\ & \leq -c_{+,k} g_k |r_{k,2}| \tau_k u_k' \hat{\tau}_{c,k}. \end{aligned} \tag{3.30}$$

Considering (3.28) and (3.30), it yields that

$$\begin{aligned} \dot{V}_2 &= \dot{V}_1 + \sum\nolimits_{k=1}^{N} r_{k,2}(h_k + f_k - \dot{\alpha}_k + g_k sat_k(dz_k(u_k))) \\ &\leq \dot{V}_1 + \sum\nolimits_{k=1}^{N} [r_{k,2} f_k + r_{k,2} h_k - r_{k,2}\dot{\alpha}_k - \\ & \quad |r_{k,2}| g_k \tau_k c_{+,k} u_k' \hat{\tau}_{c,k}]. \end{aligned} \tag{3.31}$$

(II) If $u_k \geq \frac{u_{H,k}}{c_{+,k}} + u_{+,k}$, from (3.7), we have

$$\begin{aligned} sat_k(dz_k(u_k)) = u_{H,k} &= dz_k(u_k)\frac{u_{H,k}}{dz_k(u_k)} \\ &= \omega_k(u_k)c_{+,k}(u_k - u_{+,k}), \end{aligned} \tag{3.32}$$

where

$$dz_k(u_k) = c_{+,k}(u_k - u_{+,k}), \text{ and } \omega_k(u_k) = \frac{u_{H,k}}{dz_k(u_k)}. \tag{3.33}$$

Thus, we can obtain that

$$\begin{aligned} & r_{k,2}g_k sat_k(dz_k(u_k)) \\ &= c_{+,k}g_k r_{k,2}\omega_k(u_k)(u_k - u_{+,k}) \\ &= c_{+,k}g_k r_{k,2}\omega_k(u_k)u_k'(1 + \hat{\tau}_{c,k}) \\ &= c_{+,k}g_k r_{k,2}\omega_k(u_k)u_k' + \\ &\quad c_{+,k}g_k r_{k,2}\omega_k(u_k)u_k'\hat{\tau}_{c,k}. \end{aligned} \tag{3.34}$$

Because $0 < \omega_k(dz_k(u_k)) < 1$, like doing in (3.29), one has

$$\begin{aligned} & g_k r_{k,2} sat_k(dz_k(u_k)) \\ &\leq -c_{+,k}g_k|r_{k,2}|\omega_k(dz_k(u_k))u_k'\hat{\tau}_{c,k}. \end{aligned} \tag{3.35}$$

Because $\omega_k(u_k) \geq \tau_k > 0$, we have

$$r_{k,2}g_k sat_k(dz_k(u_k)) \leq -|r_{k,2}|g_k\tau_k c_{+,k}u_k'\hat{\tau}_{c,k}. \tag{3.36}$$

Substituting (3.36) into (3.28), it yields

$$\begin{aligned} \dot{V}_2 &= \dot{V}_1 + \sum\nolimits_{k=1}^{N} r_{k,2}(h_k + f_k - \dot{\alpha}_k + g_k sat_k(dz_k(u_k))) \\ &\leq \dot{V}_1 + \sum\nolimits_{k=1}^{N} [r_{k,2}f_k + r_{k,2}h_k - r_{k,2}\dot{\alpha}_k - \\ &\quad |r_{k,2}|g_k\tau_k c_{+,k}u_k'\hat{\tau}_{c,k}]. \end{aligned} \tag{3.37}$$

Therefore, from (3.31) and (3.37), we have

$$\begin{aligned} \dot{V}_2 &= \dot{V}_1 + \sum\nolimits_{k=1}^{N} r_{k,2}(h_k + f_k - \dot{\alpha}_k + g_k sat_k(dz_k(u_k))) \\ &\leq \dot{V}_1 + \sum\nolimits_{k=1}^{N} [r_{k,2}f_k + r_{k,2}h_k - r_{k,2}\dot{\alpha}_k - \\ &\quad |r_{k,2}|g_k\tau_k c_{+,k}u_k'\hat{\tau}_{c,k}]. \end{aligned} \tag{3.38}$$

Case 2:$r_{k,2} \geq 0$
Because $r_{k,2} \geq 0$, from (3.7), we have

$$dz_k(u_k) = c_{-.k}(u_k + u_{-,k}) \text{ and } u_k = -u_k'(1 + \hat{\tau}_{c,k}) - u_{-,k} < -u_{-,k} < 0$$

Obviously, $u_k < -u_{-,k} < 0$.

Now, we give the following analysis.

(i) If $u_k > -(\frac{u_{L,k}}{c_{-,k}} + u_{-,k})$, one has

$$sat_k(dz_k(u_k)) = dz_k(u_k). \tag{3.39}$$

Since

$$dz_k(u_k) = c_{-,k}(u_k + u_{-,k}),$$

we have

$$sat_k(dz_k(u_k)) = c_{-,k}(u_k + u_{-,k}). \tag{3.40}$$

Hence, the term $g_k r_{k,2} sat_k(dz_k(u_k))$ in (3.28) can be derived as

$$\begin{aligned} & g_k r_{k,2} sat_k(dz_k(u_k)) \\ & \quad = c_{-,k} g_k r_{k,2}(u_k + u_{-,k}) \\ & \quad = -c_{-,k} g_k r_{k,2} u_k'(1 + \hat{\tau}_{c,k}) \\ & \quad = -c_{-,k} g_k r_{k,2} u_k' - c_{-,k} g_k r_{k,2} u_k' \hat{\tau}_{c,k}. \end{aligned} \tag{3.41}$$

In this work, parameter $c_{-,k}$ is positive, i.e., $c_{-,k} > 0$. Further, in this case, $r_{k,2}$ is negative, i.e., $r_{k,2} \geq 0$. From (3.17) and (3.18), we have $u_k' > 0$ and $\hat{\tau}_{c,k} > 0$. So, we have

$$\begin{aligned} & r_{k,2} = |r_{k,2}| \\ & c_{-,k} u_k' > 0, \;\; c_{-,k} u_k' \hat{\tau}_{c,k} > 0. \end{aligned} \tag{3.42}$$

From (3.41) and (3.42), it further obtains

$$\begin{aligned} & g_k r_{k,2} sat_k(dz_k(u_k)) \\ & \quad = -c_{-,k} g_k |r_{k,2}| u_k' - c_{-,k} g_k |r_{k,2}| u_k' \hat{\tau}_{c,k} \\ & \quad \leq -c_{-,k} g_k |r_{k,2}| u_k' \hat{\tau}_{c,k}. \end{aligned} \tag{3.43}$$

Substituting (3.43) into (3.28), one has

$$\begin{aligned} \dot{V}_2 &= \dot{V}_1 + \sum\nolimits_{k=1}^{N} r_{k,2}(h_k + f_k - \dot{\alpha}_k + g_k sat_k(dz_k(u_k))) \\ &\leq \dot{V}_1 + \sum\nolimits_{k=1}^{N} [r_{k,2} f_k + r_{k,2} h_k - r_{k,2}\dot{\alpha}_k - \\ & |r_{k,2}| g_k c_{-,k} u_k' \hat{\tau}_{c,k}]. \end{aligned} \tag{3.44}$$

Because $\omega_k(dz_k(u_k)) \geq \tau_k > 0$, one has

$$-|r_{k,2}|g_k c_{-,k} u'_k \hat{\tau}_{c,k} \leq -|r_{k,2}|g_k \tau_k c_{-,k} u'_k \hat{\tau}_{c,k}. \tag{3.45}$$

Substituting (3.45) into (3.44), it yields

$$\begin{aligned}\dot{V}_2 &= \dot{V}_1 + \sum\nolimits_{k=1}^{N} r_{k,2}(h_k + f_k - \dot{\alpha}_k + g_k sat_k(dz_k(u_k))) \\ &\leq \dot{V}_1 + \sum\nolimits_{k=1}^{N} [r_{k,2} f_k + r_{k,2} h_k - r_{k,2}\dot{\alpha}_k - \\ &\quad |r_{k,2}|g_k \tau_k c_{-,k} u'_k \hat{\tau}_{c,k}].\end{aligned} \tag{3.46}$$

(ii) If $u_k < -(\frac{u_{L,k}}{c_{-,k}} + u_{-,k})$, we have $dz_k(u_k) = c_{-,k}(u_k + u_{-,k})$. Obviously, $dz_k(u_k) \leq -u_{L,k}$. From (3.7), we have

$$\begin{aligned}sat_k(dz_k(u_k)) &= u_{L,k} \\ &= \omega_k(u_k) c_{-,k}(u_k + u_{-,k}),\end{aligned} \tag{3.47}$$

where $\omega_k(u_k) = \frac{u_{L,k}}{dz_k(u_k)}$.

Hence, the term $g_k r_{k,2} sat_k(dz_k(u_k))$ can be described as

$$\begin{aligned}&g_k r_{k,2} sat_k(dz_k(u_k)) \\ &\quad = c_{-,k} g_k r_{k,2} \omega_k(u_k)(u_k - u_{-,k}) \\ &\quad = -c_{-,k} g_k r_{k,2} \omega_k(u_k) u'_k (1 + \hat{\tau}_{c,k}) \\ &\quad = -c_{-,k} g_k r_{k,2} \omega_k(u_k) u'_k \\ &\qquad - c_{-,k} g_k r_{k,2} \omega_k(u_k) u'_k \hat{\tau}_{c,k}.\end{aligned} \tag{3.48}$$

Similar to (3.36), because $1 > \omega_k(u_k) > 0$, $\hat{\tau}_{c,k} > 0$, $r_{k,2} \geq 0$, $u'_k > 0$, and $c_{-,k} > 0$, one has

$$\begin{aligned}&r_{k,2} g_k sat_k(dz_k(u_k)) \\ &\quad \leq -|r_{k,2}|g_k \omega_k(dz_k(u_k)) c_{-,k} u'_k \hat{\tau}_{c,k}.\end{aligned} \tag{3.49}$$

Because $\omega_k(u_k) \geq \tau > 0$, (3.49) becomes

$$\begin{aligned}&g_k r_{k,2} sat_k(dz_k(u_k)) \\ &\quad \leq -c_{-,k} g_k |r_{k,2}| \tau_k u'_k \hat{\tau}_{c,k}.\end{aligned} \tag{3.50}$$

Substituting (3.50) into (3.28), it follows

$$\begin{aligned}\dot{V}_2 &= \dot{V}_1 + \sum\nolimits_{k=1}^{N} r_{k,2}(h_k + f_k - \dot{\alpha}_k + g_k sat_k(dz_k(u_k))) \\ &\leq \dot{V}_1 + \sum\nolimits_{k=1}^{N} [r_{k,2} f_k + r_{k,2} h_k - r_{k,2}\dot{\alpha}_k - \\ &\quad |r_{k,2}|g_k \tau_k c_{-,k} u'_k \hat{\tau}_{c,k}].\end{aligned} \tag{3.51}$$

From (3.46) and (3.51), (3.28) is derived

$$\begin{aligned}\dot{V}_2 &= \dot{V}_1 + \sum_{k=1}^{N} r_{k,2}(h_k + f_k - \dot{\alpha}_k + g_k sat_k(dz_k(u_k))) \\ &\leq \dot{V}_1 + \sum_{k=1}^{N} [r_{k,2} f_k + r_{k,2} h_k - r_{k,2}\dot{\alpha}_k - \\ &|r_{k,2}| g_k \tau_k c_{-,k} u_k' \hat{\tau}_{c,k}].\end{aligned} \tag{3.52}$$

Since $c_k = \min\{c_{+,k}, c_{-,k}\}$ and $\tau_{c,k} = g_k \tau_k c_k$, one has

$$- |r_{k,2}| g_k \tau_k c_{+,k} u_k' \hat{\tau}_{c,k} \leq -|r_{k,2}| \tau_{c,k} u_k' \hat{\tau}_{c,k}$$

$$- |r_{k,2}| g_k \tau_k c_{-,k} u_k' \hat{\tau}_{c,k} \leq -|r_{k,2}| \tau_{c,k} u_k' \hat{\tau}_{c,k}.$$

From (3.38) and (3.52), one has

$$\begin{aligned}\dot{V}_2 &= \dot{V}_1 + \sum_{k=1}^{N} r_{k,2}(h_k + f_k - \dot{\alpha}_k + g_k sat_k(dz_k(u_k))) \\ &\leq \dot{V}_1 + \sum_{k=1}^{N} [r_{k,2} f_k + r_{k,2} h_k - r_{k,2}\dot{\alpha}_k - |r_{k,2}| \tau_{c,k} u_k' \hat{\tau}_{c,k}].\end{aligned} \tag{3.53}$$

From (3.16), it is easy to find that $\dot{\alpha}_k$ is continuous. From the description below (3.1), $f_k(\varpi_k)$ also is continuous. Hence, define a new function $f_k'(\bar{x}_k) = f_k(\varpi_k) - \dot{\alpha}_k$. Obviously, $f_k'(\bar{x}_k)$ is continuous, which means that there exists a fuzzy logic system such that the unknown function $f_k'(\bar{x}_k)$ is described as

$$f_k'(\bar{x}_k) = \theta_k^{*T} S_k(\bar{x}_k) + \varepsilon_k(\bar{x}_k),$$

where $\bar{x}_k = [x_{k,1}, \ldots, x_{k,m_k}]^T = [1, r_j, \pi_k, \varpi_k, \pi_j]^T$, $j \in N_k$. So, (3.53) is developed as

$$\dot{V}_2 \leq \dot{V}_1 + \sum_{k=1}^{N} [r_{k,2}\theta_k^{*T} S_k(\bar{x}_k) + r_{k,2}(h_k + \varepsilon_k) - |r_{k,2}| \tau_{c,k} u_k' \hat{\tau}_{c,k}]. \tag{3.54}$$

Further, it follows from Assumptions 3.2 and 3.3 that $|h_k| \leq M_{h_k}$ and $|\varepsilon_k(\bar{x}_k)| \leq M_{\varepsilon k}$. Hence,

$$r_{k,2}(h_k + \varepsilon_k) \leq |r_{k,2}| M_{h\varepsilon k}$$

where $M_{h\varepsilon k} = M_{h_k} + M_{\varepsilon k}$. Thus, (3.54) is developed as

$$\begin{aligned}\dot{V}_2 \leq \dot{V}_1 &+ \sum_{k=1}^{N} [r_{k,2}\theta_k^{*T} S_k(\bar{x}_k) + |r_{k,2}| M_{h\varepsilon k} - \\ &|r_{k,2}| \tau_{c,k} u_k' \hat{\tau}_{c,k}].\end{aligned} \tag{3.55}$$

Now, $r_{k,2}\hat{\theta}_k^T S_k(\bar{x}_k)$ is added to and subtracted from the right side of (3.55). Then, one has

$$\begin{aligned}\dot{V}_2 \le& \dot{V}_1 + \sum_{k=1}^{N}[-r_{k,2}\tilde{\theta}_k^T S_k(\bar{x}_k) + r_{k,2}\hat{\theta}_k^T S_k(\bar{x}_k)+ \\ &|r_{k,2}|M_{h\varepsilon k} - |r_{k,2}|g_k\tau_k c_{+,k}u_k'\hat{\tau}_{c,k}] \\ \le& \dot{V}_1 + \sum_{k=1}^{N}[-r_{k,2}\tilde{\theta}_k^T S_k(\bar{x}_k) + |r_{k,2}||\hat{\theta}_k^T S_k(\bar{x}_k)|+ \\ &|r_{k,2}|M_{h\varepsilon k} - |r_{k,2}|\tau_{c,k}u_k'\hat{\tau}_{c,k}],\end{aligned} \tag{3.56}$$

where $\tilde{\theta}_k = \hat{\theta}_k - \theta_k^*$.

Substituting (3.27) into (3.56), it yields

$$\begin{aligned}\dot{V}_2 \le& \sum_{k=1}^{N}[-(l_k - \frac{d_k + b_k}{2})r_{k,1}^2 + \frac{(d_k + b_k)}{2}r_{i,2}^2 - \\ &r_{k,2}\tilde{\theta}_k^T S_k + |r_{k,2}||\hat{\theta}_k^T S_k| + |r_{k,2}|M_{h\varepsilon k} - \\ &|r_{k,2}|\tau_{c,k}u_k'\hat{\tau}_{c,k}],\end{aligned} \tag{3.57}$$

where S_k denotes $S_k(\bar{x}_k)$.

Define

$$V = V_2 + \sum_{k=1}^{N}(\frac{\tilde{\tau}_{c,k}^2}{2\xi_\tau} + \frac{\tilde{M}_{h\varepsilon k}^2}{2\xi_M} + \frac{\tilde{W}_k^T\tilde{\theta}_k}{2\xi_\theta}),$$

where

$$\tilde{\tau}_{c,k} = \frac{1}{\hat{\tau}_{c,k}} - \tau_{c,k} = \hat{\tau}_{c,k}^{-1} - \tau_{c,k},$$

$$\tilde{M}_{h\varepsilon k} = \hat{M}_{h\varepsilon k} - M_{h\varepsilon k}, \quad \tilde{W}_k = \hat{W}_k - W_k^*.$$

For V, its time derivative is

$$\dot{V} = \dot{V}_2 + \sum_{k=1}^{N}[\tilde{\tau}_{c,k}\dot{\tilde{\tau}}_{c,k}/\xi_\tau + \tilde{M}_{h\varepsilon k}\dot{\hat{M}}_{h\varepsilon k}/\xi_M + \tilde{\theta}_k^T\dot{\hat{\theta}}_k/\xi_\theta] \tag{3.58}$$

Substituting (3.57) into (3.58), we have

$$\begin{aligned}\dot{V} \le& \sum_{k=1}^{N}[-(l_k - \frac{b_k + d_k}{2})r_{k,1}^2 + \frac{(b_k + d_k)}{2}r_{i,2}^2 - \\ &r_{k,2}\tilde{\theta}_k^T S_k(\bar{Z}_k) + |r_{k,2}||\hat{\theta}_k^T S_k(\bar{Z}_k)| + |r_{k,2}|M_{h\varepsilon k} - \\ &|r_{k,2}|\tau_{c,k}u_k'\hat{\tau}_{c,k}] + \\ &\tilde{\tau}_{c,k}\dot{\tilde{\tau}}_{ck}/\xi_\tau + \tilde{M}_{h\varepsilon k}\dot{\hat{M}}_{h\varepsilon k}/\xi_M + \tilde{\theta}_k^T\dot{\hat{\theta}}_k/\xi_\theta.\end{aligned} \tag{3.59}$$

Since $\tilde{\tau}_{c,k} = \frac{1}{\hat{\tau}_{c,k}} - \tau_{c,k} = \hat{\tau}_{c,k}^{-1} - \tau_{c,k}$, we have

$$\dot{\tilde{\tau}}_{c,k} = d(\hat{\tau}_{c,k}^{-1})/dt - \dot{\tau}_{c,k} = -\hat{\tau}_{c,k}^{-2}\dot{\hat{\tau}}_{c,k} \tag{3.60}$$

$$\begin{aligned}
&-|r_{k,2}|\tau_{c,k}u'_k\hat{\tau}_{c,k} \\
&\quad = |r_{k,2}|(\hat{\tau}_{c,k}^{-1} - \tau_{c,k})u'_k\hat{\tau}_{c,k} - |r_{k,2}|\hat{\tau}_{c,k}^{-1}u'_k\hat{\tau}_{c,k} \\
&\quad = |r_{k,2}|\tilde{\tau}_{c,k}u'_k\hat{\tau}_{c,k} - |r_{k,2}|u'_k.
\end{aligned} \tag{3.61}$$

Substituting (3.60) and (3.61) into (3.59), it yields

$$\begin{aligned}
\dot{V} \leq & \sum\nolimits_{k=1}^{N}[-(l_k - \frac{b_k + d_k}{2})r_{k,1}^2 + \frac{(b_k + d_k)}{2}z_{i,2}^2 - \\
& r_{k,2}\tilde{\theta}_k^T S_k(\bar{Z}_k) + |r_{k,2}||\hat{\theta}_k^T S_k(\bar{Z}_k)| + |r_{k,2}|M_{h\varepsilon k} + \\
& |r_{k,2}|\tilde{\tau}_{c,k}u'_k\hat{\tau}_{c,k} - |r_{k,2}|u'_k - \\
& \hat{\tau}_{c,k}^{-2}\tilde{\tau}_{c,k}\dot{\hat{\tau}}_{ck}/\xi_\tau + \tilde{M}_{h\varepsilon k}\dot{\hat{M}}_{h\varepsilon k}/\xi_M + \tilde{W}_k^T\dot{\hat{\theta}}_k/\xi_\theta.
\end{aligned} \tag{3.62}$$

Let us recall the adaptive laws (3.19)–(3.21):

$$\dot{\hat{\tau}}_{c,k} = \xi_\tau \hat{\tau}_{c,k}^3 u'_k |r_{k,2}|, \quad \hat{\tau}_{c,k}(0) > 0$$

$$\dot{\hat{M}}_{h\varepsilon k} = \xi_M |r_{k,2}|, \quad \hat{M}_{h\varepsilon k}(0) > 0$$

$$\dot{\hat{\theta}}_k = \xi_\theta r_{k,2} S_k, \quad \hat{\theta}_k(0) > 0,$$

where parameters ξ_θ, ξ_M, and ξ_τ are positive real numbers.

Substituting the above adaptive laws defined in (3.18)–(3.21) into (3.62), it results

$$\dot{V} \leq \sum\nolimits_{k=1}^{N}[-(l_k - \frac{d_k + b_k}{2})r_{k,1}^2 - r_{i,2}^2] \leq 0. \tag{3.63}$$

From (3.63), we can find, V is non-increasing monotonically, which means that $V(+\infty)$ also exists. Based Lyapunov theory, one obtains $\lim_{t\to\infty} r_{k,1}(t) = 0$. Finally, it is obtained that

$$\text{For } k = 1, 2, \ldots, N, \lim_{t\to\infty} |e_k(t)| = 0.$$

Because $e = [e_1, \ldots, e_N]'$ and

$$r_{k,1} = e_k - \sum\nolimits_{j\in N_k}[a_{kj}(\pi_j - \pi_k) - b_k(\pi_0 - \pi_k)],$$

one has

$$\lim_{t\to\infty} e(t) = \mathbf{0}.$$

Further, it follows from Lemma 3.3 that

$$||\delta|| \leq \frac{||e||}{\underline{\sigma}(L + B)}.$$

Therefore, the following results are obtained

$$\lim_{t\to\infty}\delta(t)=\mathbf{0} \text{ and } \lim_{t\to\infty}\delta_k(t)=0,$$

which means that the output synchronization between the followers and leader has been achieved, and further means each output tracking error synchronizes asymptotically to the origin. The proof is completed here.

3.4 Simulation

In this simulation, the leader-following system consists of one leader and three followers. In the system, each node is an autonomous underwater vehicle (AUV) (Wang et al. 2010; Chao et al. 2011). In order to verify the results obtained in this chapter, their steering consensus problems are considered.

3.4.1 Descriptions of AUVs

In this chapter, the AUVs are described as (Wang et al. 2010; Chao et al. 2011)

$$\begin{cases} M(v)\dot{v}+C(v)v+D(v)v+g(\tau)=\tau \\ \dot{\tau}=J(\tau)v, \end{cases} \tag{3.64}$$

where $\tau=[\tau_1^T,\tau_2^T]^T=[x,y,z,\phi,\theta,\pi]^T$ and $v=[v_1^T,v_2^T]^T=[u,\upsilon,w,p,q,r]^T$, inertial matrix $M(v)\in R^{6\times6}$, $C(v)\in R^{6\times6}$, damping matrix $D(v)\in R^{6\times6}$, $g(\tau)$ denotes gravitational force and moments, control input $\tau\in R^6$. For the details of AUVs, it can found in Hong et al. (2008), Yoon and Qiao (2011), Wang et al. (2010), Chao et al. (2011).

From the vehicle's kinematic, it is derived as

$$\dot{\tau}=J(\tau_2)v=\begin{pmatrix} J_1(\tau_2) & 0_{3\times3} \\ 0_{3\times3} & J_2(\tau_2)\end{pmatrix}\begin{pmatrix} v_1 \\ v_2\end{pmatrix} \tag{3.65}$$

where

$$J_1(\tau_2)=\left(\begin{matrix}\cos(\theta)\cos(\pi) \\ \cos(\theta)\sin(\pi) \\ -\sin(\theta)\end{matrix}\right. \rightarrow$$

$$\begin{matrix}\sin(\phi)\sin(\theta)\cos(\pi)-\cos(\phi)\sin(\pi) \\ \sin(\phi)\sin(\theta)\sin(\pi)+\cos(\phi)\cos(\pi) \\ \sin(\phi)\cos(\theta)\end{matrix} \rightarrow$$

$$\left.\begin{matrix} \sin(\phi)\sin(\pi)+\cos(\phi)\sin(\theta)\cos(\pi) \\ \cos(\phi)\sin(\theta)\sin(\pi)-\sin(\phi)\cos(\pi) \\ \cos(\phi)\cos(\theta) \end{matrix}\right)$$

$$J_2(\tau_2) = \begin{pmatrix} 1 & \sin(\phi)\tan(\theta) & \cos(\phi)\tan(\theta) \\ 0 & \cos(\phi) & -\sin(\phi) \\ 0 & \frac{\sin(\phi)}{cos(\theta)} & \frac{\cos(\phi)}{cos(\theta)} \end{pmatrix}.$$

For each AUV, it is divided into three subsystems (Chao et al. 2011): (1) diving control subsystem, (2) speed control subsystem, and (3) steering control subsystem. For the steering control subsystem, it can be simplified (Wang et al. 2010; Chao et al. 2011; Wang 2014):

$$\begin{cases} \dot{\pi} = r \\ \dot{\varpi} = f(r) + h(t) + g\delta_r, \end{cases} \tag{3.66}$$

where r and π denote, respectively, yaw velocity and steering angle; rudder deflection is denoted by δ_r; disturbance, uncertainty, and modeling errors are integrated and represented together as $h(t)$, which is bounded. Control gain $g > 0$ is a real number.

In order to distinguish the AUVs, an additional subscript k $(k = 1, \ldots, N)$ is added to the symbols in (3). Thus, for AUV k, it is described as

$$\begin{cases} \dot{\pi}_k = \varpi_k \\ \dot{\varpi}_k = f_k(\varpi_k) + h_k + g_k\delta_{rk}, \end{cases} \tag{3.67}$$

where δ_{rk}, ϖ_k, and π_k, respectively, denote the rudder deflection, yaw velocity, and steering angle; g_k is gain constant.

The AUV, as a leader, is labeled 0 and has the following form:

$$\begin{cases} \dot{\pi}_0 = r_0 \\ \dot{\varpi}_0 = f_0(r_0, t), \end{cases} \tag{3.68}$$

where $r_0 \in R$ and π_0 denote the yaw velocity and steering angle, $f_0(r_0, t)$ is unknown to the other AUVs.

Here, multiple autonomous underwater vehicles are considered and described by graph $\bar{G}$, communication topology of which is shown in Fig. 3.4.

3.4.2 Simulation Results

By choosing the parameters in Wang et al. (2010), Chao et al. (2011), Wang (2014), steering subsystem is described as

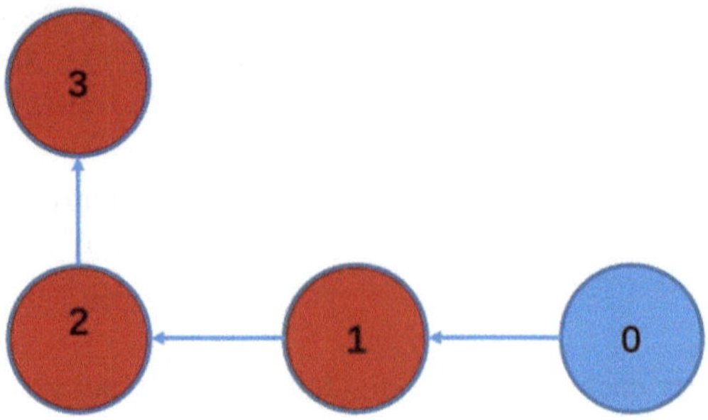

Fig. 3.4 Communication topology $\bar{G}$

$$\begin{cases} \dot{\psi}_k = \varpi_k \\ \dot{\varpi}_k = -1.5380\varpi_k|\varpi_k| - 3.1514v_k|v_k| + 0.1255\varpi_k - 5.9330\delta_{rk}. \end{cases} \tag{3.69}$$

As doing in Wang et al. (2010), Chao et al. (2011), Wang (2014), sway angle v_k is chosen as $v_k = 0.1sin(t) - 3.1514v_k|v_k|$, which is taken as disturbance h_k. In this work, $\delta_{r0} = 0$.

For the saturation and dead-zone, the parameters are set as $c_{-,k} = 0.9$, $c_{+,k} = 0.85$, $u_{-,k} = 0.15$, $u_{+,k} = 0.1$, $u_{L,k} = 5.5$, $u_{H,k} = 5$.

The initial condition is $\psi_0 = 0^0$, $r_0 = 0^0$, $\psi_1 = -3.6^0$, $r_1 = 0^0$, $\psi_2 = -2.7^0$, $r_2 = 0^0$, $\psi_3 = -1.8^0$, $r_3 = 0^0$. For neural networks, their weight $\hat{W}_k \in R^{10}$ is randomly taken in interval (0, 1]. The design parameters are, respectively, set as $\xi_M = 1$, $l = 3$, $\xi_W = 1$, $\xi_\tau = 1$,. From Fig. 3.1, one can obtain that $B = diag\{1, 0, 0\}$.

Figures 3.5, 3.6, 3.7, 3.8, and 3.9 show simulation results. Under the FTC (3.26), the dead-zone and saturation in the controllers are compensated, and their effects have been reduced. Thus, for steering angles and yaw velocities, the followers can asymptotically synchronize to the leader shown in Figs. 3.5 and 3.6. Figures 3.7 and 3.8 show that the synchronization errors between the leader and followers converge

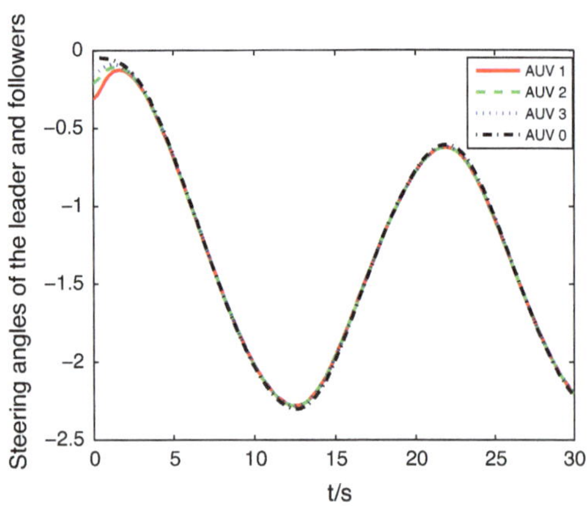

Fig. 3.5 The steering angles

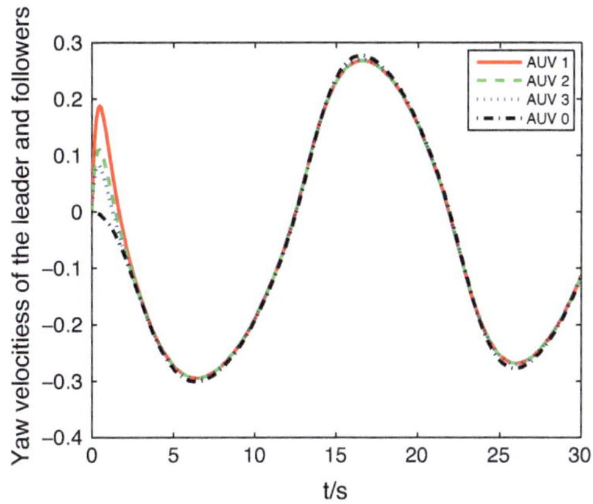

Fig. 3.6 The yaw velocities

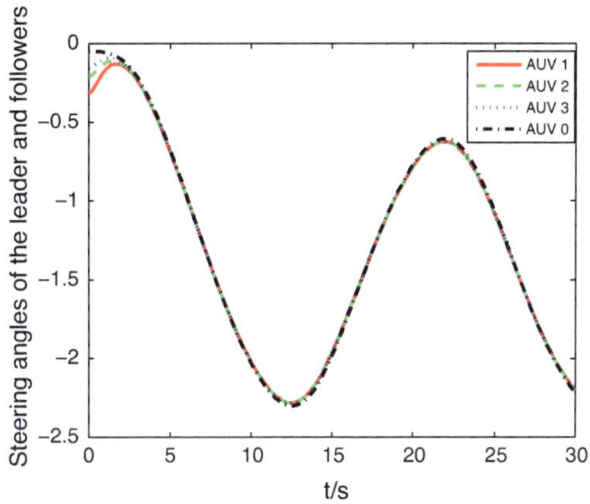

Fig. 3.7 The synchronization errors of steering angles

to an adjustable neighborhood of the origin. The boundedness of input signals is shown in Fig. 3.9.

3.5 Conclusion

This chapter studies the output synchronization of a class of leader-following systems subjected simultaneously to dead-zone and saturation. By using the approximation capability of fuzzy logic systems, combining adaptive control and backstepping control technology, a distributed fault-tolerant control strategy is constructed to ensure that the outputs' synchronization between the leader and followers is achieved. Simulation results of multiple AUVs demonstrate the effectiveness of the strategy proposed in this work.

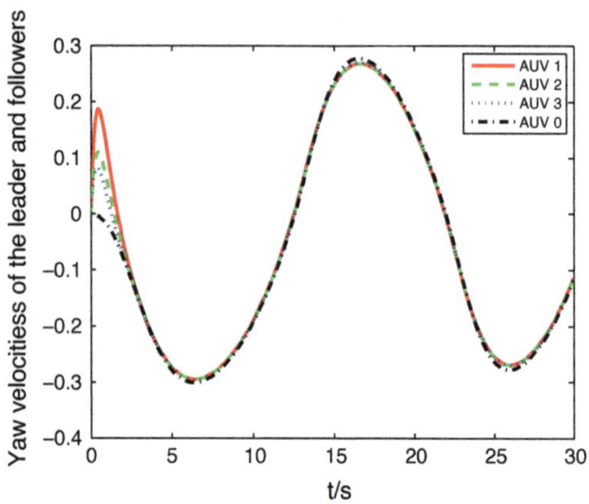

Fig. 3.8 The synchronization errors of yaw velocities

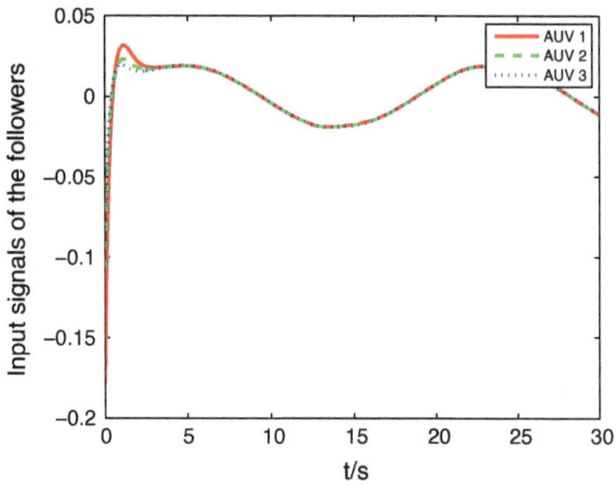

Fig. 3.9 Input signals

References

Ahn CK (2013) $l_2 - l_\infty$ elimination of overflow oscillations in 2-D digital filters described by user model with external interference. IEEE Trans Circuits Syst II: Express Briefs 60(6):361–365

Ahn CK (2014) $l_2 - l_\infty$ suppression of limit cycles in interfered two-dimensional digital filters: a Fornasini-CMarchesini model case. IEEE Trans Circuits Syst II: Express Briefs 61(8):614–618

Ahn CK, Kar H (2015a) Expected power bound for two-dimensional digital filters in the fornasini-marchesini local state-space model. IEEE Signal Process Lett 22(8):1065–1069

Ahn CK, Kar H (2015b) Passivity and finite-gain performance for two-dimensional digital filters: the FM LSS model case. IEEE Trans Circuits Syst II: Express Briefs 62(9):871–875

Chao SZ, Hong GS, Zhao EY, Mandar, C (2011) Modular modeling of autonomous underwater vehicle. Oceans. IEEE, pp 1–6

Chen J, Cao X, Cheng P, Xiao Y, Sun Y (2010) Distributed collaborative control for industrial automation with wireless sensor and actuator networks. IEEE Trans Ind Electron 57(12):4219–4230

Cho HY, Bai EW (1998) Convergence results for an adaptive dead zone inverse. Int J Adapt Control Signal Proc 12(5):451–466

Fiorelli E, Leonard NE, Bhatta P (2006) Multi-AUV control and adaptive sampling in monterey bay. IEEE J Ocean Eng 31(4):935–948
Hong Y, Chen G, Bushnell L (2008) Distributed observers design for leader-following control of multi-agent networks. Automatica 44(3):846–850
Hsu KC, Wang WY, Lin PZ (2004) Sliding mode control for uncertain nonlinear systems with multiple inputs containing sector nonlinearities and deadzones. IEEE Trans Syst Man Cybern Part B: Cybern 34(1):374–380
Olfati-Saber R, Fax J, Murray R (2007) Consensus and cooperation in networked multi-agent systems. Proc IEEE 95(1):215–233
Qu Z (2009) Cooperative control of dynamical system: application to autonomous vehicles. Springer, London
Ren W, Moore KL, Chen Y (2007) High-order and model reference consensus algorithms in cooperative control of multivehicle systems. ASME J Dyn Sys Meas Control 129(5):678–688
Ren W, Cao C (2011) Distributed coordination of multi-agent networks: emergent problems, models, and issues. Springer, London
Shen Q, Shi P (2015) Distributed command filtered backstepping consensus tracking control of nonlinear multiple-agent systems in strict-feedback form. Automatica 53(3):120–124
Shen Q, Shi P (2016a) Distributed adaptive fuzzy Control for nonlinear multi-agent systems via sliding mode observers. IEEE Trans Cybern 46(12):3086–3097
Shen Q, Shi P (2016b) Output consensus control of multi-agent systems with unknown nonlinear dead-zone. IEEE Trans Syst Man Cybern: Syst 46(10):1329–1337
Shen QK, Jiang B, Shi P (2014) Cooperative adaptive fuzzy tracking control for networked unknown nonlinear multi-agent systems with time-varying actuator faults. IEEE Trans Fuzzy Syst 22(3):494–504
Shen Q, Shi P, Zhu J, Wang S, Shi Y (2020) Neural networks based distributed adaptive control of nonlinear multi-agent systems. IEEE Trans Neural Netwo Learn Syst 31(3):1010–1021
Shi P, Shen QK (2015) Cooperative control of multi-agent systems with unknown state-dependent controlling effects. IEEE Trans Autom Sci Eng 12(3):827–834
Shyu KK, Liu WJ, Hsu KC (2005) Design of large-scale time-delayed with dead-zone input via variable structure control. Automatica 41(7):1239–1246
Tao G, Kokotovic PV (1994) Adaptive sliding control of plants with unknown dead-zone. IEEE Trans Autom Control 39(1):59–68
Tao G, Kokotovic PV (1995) Discrete-time adaptive control of systems with unknown dead-zone. Int J Control 61(1):1–17
Wang J (2014) Nonsmooth PD-like finite time steering control for autonomous underwater vehicles. J Syst Sci Math Sci 32(12):1527–1531
Wang XS, Su C-Y, Hong H (2004) Robust adaptive control a class of nonlinear systems with an unknown dead-zone. Automatica 40(3):407–413
Wang J, Song X, Zhao Y, Zhang G (2010) Nonlinear switched diving control for under-actuated autonomous underwater vehicles. In: Proceedings of the 29th Chinese control conference, pp 705–710
Wang W, Yu Y (2016) Fuzzy adaptive consensus of second-order nonlinear multi-agent systems in the presence of input saturation. Int J Innov Comput Inf Control (12)2:533–542
Yan JJ, Shyu KK, Lin JS (2005) Adaptive variable structure control for uncertain chaotic systems containing dead-zone nonlinearity. Chaos Solitions Fract 25(2):347–355
Yau HT, Yan JJ (2004) Design of sliding mode controller for Lorenz chaotic system with nonlinear input. Chaos Solitions Fract 19(4):891–898
Yoon S, Qiao CM (2011) Cooperative search and survey using autonomous underwater vehicles. IEEE Trans Parallel Distrib Syst 22(3):364–379
Yu H, Xia X (2012) Adaptive consensus of multi-agents in networks with jointly connected topologies. Automatica 48(8):1783–1790
Zhang H, Lewis FL (2012) Adaptive cooperative tracking control of higher-order nonlinear systems with unknown dynamics. Automatica 48(7):1432–1439

Zhang M, Shen X, Li T (2016) Fault tolerant attitude control for cubeSats with input saturation based on dynamic adaptive neural network. Int J Innov Comput Inf Control 12(2):651–663
Zhu W, Cheng D (2010) Leader-following consensus of second-order agents with multiple time-varying delays. Automatica 46(12):1994–1999

Chapter 4
Distributed Neural Fault-Tolerant Supervisory Control Against Backlash-Like Hysteresis

4.1 Introduction

Neural networks (NNs) and fuzzy logic systems (FLSs) (Wang 1993, 1994; Shen et al. 2019; Su et al. 2000), as two universal function approximations, are utilized widely in the control design of nonlinear systems. For the approximation theories of NNs and FLSs, it is under the necessary condition that they are able to approximate a continuous unknown function as precisely as possible. The necessary condition is that all the variables in the function must always be within a compact set. In fact, the boundedness of approximation error is guaranteed by the condition. Now, let's take function $f(y)$ as an example to illustrate clearly it, where y can be a scalar or vector. It can be approximated by NNs or FLSs. Without losing the generality, its estimate is denoted by $\hat{f}(y)$. Obviously, approximation error can be described as $\varepsilon(y) = f(y) - \hat{f}(y)$. There exists a compact set Ω_y is a compact set related to y. For the function $f(y)$, only if y or each element of y always belongs to the set Ω_y, it can be accurately approximated by neural networks or fuzzy logic systems, which implies that $\varepsilon(y)$ is bounded on the set. In other words, if the condition about the boundedness of y is not met, then the function $f(y)$ cannot be approximated with arbitrary precision by neural networks or fuzzy logic system, because the error $\varepsilon(y) = f(y) - \hat{f}(y)$ is not always bounded and cannot be further compensated. When approximating unknown functions by using neural networks or fuzzy logic systems in control designing, the first task is to ensure that the above condition is met. Recently, by using NNs or FLSs, there are abundant control methods have been obtained for leader-following systems (LFSs). Among the results (Shen et al. 2013, 2014), however, for the mentioned-above condition, no clear and suitable method has been given to guarantee that the condition is met.

In practical applications, supervisor control strategy often be utilized to guarantee that the above condition is met (Zhang 2002a, b).

Q. Shen, *Distributed Fault-Tolerant Consensus Control of Leader-Following Systems*, Intelligent Control and Learning Systems 11,
https://doi.org/10.1007/978-981-99-7426-9_4

In the strategy, the condition about the boundedness is first ensured. Next, the main task is to construct corresponding control schemes for control objectives. Meanwhile, adaptive compensation terms are designed to compensate for the optimal approximation errors, which removes or reduces the effects of the terms. However, as far as I know, supervisor control is mainly applied to single-agent systems, and multi-agent systems do not yet involve supervisor control strategies. In fact, up to now, for MASs and LFSs, there are no reports about supervisor control methods yet. For NNs-based and FLSs-based control methods, if supervision control is not used to ensure that the above-mentioned condition about the state boundedness is true, strictly speaking, these control methods are imperfect or even wrong. Therefore, for complex systems, including the multi-agent systems studied in this chapter, designing supervisory control strategies remains an open challenging issue.

For MASs and LFSs, as the research progressed, multiple constraints were added to the system models, which makes control design more complex and difficult (Shen and Shi 2016; Shen et al. 2016, 2017, 2019). Among the constraints, input nonlinearities such as dead-zone (Shen and Shi 2016) and sector nonlinear input (Shen et al. 2017) are common. In fact, actuator failure can also cause the nonlinearities. In order to accurately control the system to achieve satisfactory control performance, these nonlinearities must be considered in control design. Backlash-like hysteresis, as a common input nonlinearity, was considered in the control design of single-agent systems (Su et al. 2000; Tao and Kokotovic 1995; Ikhouane 2020; Chwastek and Gozdur 2019; Tao et al. 2019; Elbohy et al. 2019; Starodubtsev et al. 2019; Park et al. 2019). In fact, the hysteresis that occurs in multi-agent systems has a greater negative impact on the entire system. Therefore, for MASs including LFSs, how to compensate for the hysteresis is very necessary and urgent.

In stability analysis and control designing, sliding mode control (SMC) is a powerful tool. However, signum function often occurs in the control laws by using SMC (Defoort et al. 2008), which means that the control laws are non-continuous. However, the backlash-like hysteresis considered in this work has a property, namely, the input must be continuous. Hence, the results in Defoort et al. (2008), Shen et al. (2014) can not be extended directly to leader-following systems and the other MASs.

In this chapter, for a class of leader-following systems, with backlash-like hysteresis taken into account, synchronization between the leader and followers is studied. The main contributions in this work are worth to be highlighted. (I) Backlash-like hysteresis is considered and compensated, and unknown control gain functions, which are dependent on follower state, are investigated; (II) In order to guarantee the continuities of control laws of followers, a new control scheme is constructed; and (III) Control scheme with supervisor is extended firstly to the leader-following systems, where the boundedness of follower states is ensured and theoretical analysis and control design are more rigorous and complete.

4.2 Problem Formulation and Preliminaries

In this section, basic graph theory is introduced first. Next, the consensus control problem is formulated for a class of leader-following systems. Then, some valuable lemmas are given. Backlash-like hysteresis also is introduced. Finally, neural networks are briefly described.

4.2.1 *Basic Graph Theory*

As stated in Chap. 3, in the sense of mathematics, MASs including LFSs investigated in this book can be represented by a graph. Obviously, it involves basic graph theory in the research of MASs. So, let us first introduce it in this subsection. For basic graph theory, it was already introduced in the previous chapter. For the convenience of reading, its related knowledge will be re-introduced here again. If the readers have gained an understanding of the basic graph theory, the readers can skip this part and read the following content directly.

A class of multi-agent systems will be considered here, which consists of N followers and a leader. For the leader and followers, the information exchange among the leader and followers is often modeled as a graph. Hence, we first introduce the knowledge of basic graph theory in this subsection.

A directed graph can be represented as $\mathscr{G} = (\mathscr{V}, \mathscr{E})$, where agent set $\mathscr{V}$ and edge (arcs) set $\mathscr{E}$ are defined as $\mathscr{V} = (\mathscr{V}_1, \ldots, \mathscr{V}_N)$ and $\mathscr{E} \subseteq \mathscr{V} \times \mathscr{V}$, respectively. Note that, if there exists an edge from node i to node j, then $(\mathscr{V}_i, \mathscr{V}_j) \in E$. For $\mathscr{G}$, its adjacency matrix $\mathscr{A}$ can be described as $\mathscr{A} = [a_{ij}] \in R^{N\times N}$, where element a_{ij} is defined as follows: (1) $a_{ij} = 0$ if there exists no an edge from node i to node j, i.e., $(\mathscr{V}_i, \mathscr{V}_j) \notin \mathscr{E}$; and (2) $a_{ij} \neq 0$ if there exists no an edge from node i to node j, i.e., $(\mathscr{V}_i, \mathscr{V}_j) \in E$. Throughout this chapter, a_{ii} is assumed to be equal to zero, i.e., $a_{ii} = 0$, and the topology is also assumed to be fixed, which means that $\mathscr{A}$ is fixed and is not variant.

In directed graph $\mathscr{G}$, for the i-th agent, its in-degree d_i is defined as $d_i = \sum_{j=1}^{N} a_{ij}$. Correspondingly, for the directed graph $\mathscr{G}$, its in-degree matrix and Laplacian matrix, i.e., $\mathscr{D}$ and $\mathscr{L}$, are also defined, respectively, as $\mathscr{D} = diag\{d_1, \ldots, d_N\} \in R^{N\times N}$ and $\mathscr{L} = [l_{ij}] = \mathscr{D} - \mathscr{A} \in R^{N\times N}$. For the Laplacian matrix $\mathscr{L}$, $L\underline{1} = 0$ with $\underline{1} = [1, \ldots, 1]^T \in R^{N\times 1}$. In addition, for agent i, its neighbor set can be described as $N_i = \{j|(\mathscr{V}_j, \mathscr{V}_i) \in E\}$. Obviously, for the i-th agent, if the j-th agent is its neighbor, agent i can obtain the information from the j-th agent. For directed graph, however, not necessarily vice versa. For undirected graph, neighbors are interrelated. From node i to node j, if there exists a sequence of successive arcs, which can be described in the form $\{(\mathscr{V}_i, \mathscr{V}_l), (\mathscr{V}_l, \mathscr{V}_k), \ldots, (\mathscr{V}_m, \mathscr{V}_j)\}$, then a direct path exists. In a digraph, if there exist a special node and a directed path from the special node to every other node, then graph is said to has a spanning tree. For any two different nodes $\mathscr{V}_i$ and

$\mathscr{V}_j$ in the digraph, from $\mathscr{V}_i$ to $\mathscr{V}_j$, if there always exists a direct path, then the graph is said to be strongly connected.

Let $\bar{\mathscr{E}} \subseteq \bar{\mathscr{V}} \times \bar{\mathscr{V}}$ and $\bar{\mathscr{V}} = \{\mathscr{V}_0, \mathscr{V}_1, \mathscr{V}_2, \ldots, \mathscr{V}_N\}$. Then, based on the graph $\mathscr{G}$ and a leader node, a new graph $\bar{\mathscr{G}}$ is defined as $\bar{\mathscr{G}} = \{\bar{\mathscr{V}}, \bar{\mathscr{E}}\}$. Note that, from the leader node to node k, if there exists an edge, then its weight $b_k > 0$, otherwise $b_k = 0$, where $k = 1, 2, \ldots, N$.

Now, the following lemma and assumption are made.

Lemma 4.1 (Zhang and Lewis (Shen and Shi 2016; Shen et al. 2016, 2017, 2019)) *Define* $q = (\mathscr{B} + \mathscr{L})^{-1}\underline{1} = [q_1, \ldots, q_N]$,

$$P = diag\{\frac{1}{q_i}\} \in R^{N \times N} \ \ with \ \ p_i = \frac{1}{q_i}, \quad Q = P(\mathscr{B} + \mathscr{L}) + (\mathscr{B} + \mathscr{L})^T P,$$

then matrices Q *and* P *are positive definite, i.e.,* $Q > 0$ *and* $P > 0$.

Assumption 4.1 There exists a spanning tree in graph $\bar{\mathscr{G}}$, where the leader is the root, and its topology is invariant.

4.2.2 *Problem Formulation*

In this subsection, a multi-agent system is investigated, which consists of $N + 1$ nodes, which are, respectively, labeled $0, 1, \ldots, N$. Node 0 denotes a leader, and nodes k, $k = 0, 1, 2, \ldots, N$, denote the followers. The kth follower is described as follows:

$$\begin{cases} \dot{z}_{k,i} = z_{k,i+1}, \quad i = 1, \ldots, n-1 \\ \dot{z}_{k,n} = f_k(z_k) + g_{0,k}(z_k) w_k(u_k) + \varsigma_k(x_k, t) \\ y_k = z_{k,1} \end{cases} \tag{4.1}$$

where $k = 1, \ldots, N$, $z_k = [z_{k,1}, \ldots, z_{k,n}]^T \in R^n$ is state vector, $u_k \in R$ is input and $y_k \in R$ denote output; function $f_k(\cdot) \in R$ and gain $g_{0,k}(\cdot) \in R$ are unknown but continuous; external disturbance $\varsigma_k(x_k, t) \in R$ is bounded; for the backlash-like hysteresis H considered in this work, $w_k(u_k)$ denotes its output, which will be described in the following subsection.

In the MASs, node 0, i.e., leader node, can be represented as

$$\begin{cases} \dot{z}_{0,i} = z_{0,i+1}, \quad i = 1, \ldots, n-1 \\ \dot{z}_{0,2n} = f_0(\bar{z}_0, t) \end{cases} \tag{4.2}$$

where state vector $z_0 = [z_{0,1}, z_{0,2}, \ldots, z_{0,n}]^T \in R^n$, $f_0(z_0, t) \in R$ is bounded, which is unknown for all the followers.

Note that, all the nodes including the followers and leader are modeled in a weighted graph $\bar{G}$, where $\bar{v} = \{v_0, v_1, \ldots, v_N\}$ and $\bar{E} \subseteq \bar{v} \times \bar{v}$. For the graph, the detailed description can be found in the previous section.

For the kth follower, its tracking error is defined as

$$\varpi_{k,i} = z_{k,i} - z_{0,i}, \quad k = 1, \ldots, N, \quad i = 1, \ldots, n$$

The main task in this work is, to design a control strategy with supervisor such that (1) the boundedness of the followers' states is guaranteed; and (2) the consensus between the leader and followers is achieved.

To design distributed controller for each follower, the following assumptions and lemmas are given.

Assumption 4.2 Function $f_0(z_0, t)$ is bounded, namely, $\exists$ unknown constant $M_{f_0} > 0 \in R$, $\ni |f_0(z_0, t)| \leq M_{f_0}$.

Assumption 4.3 For $k = 1, 2, \ldots, N$, $\exists$ continuous known functions $F_k(z_k) > 0 \in R \ni F_k(z_k) > f_k(z_k, t)$.

Assumption 4.4 For $k = 1, 2, \ldots, N$, functions $g_{0,k}(z_k)$ are bounded, namely, $\exists$ known constants $g_{k,1} > 0 \in R$ and $g_{k,2} > 0 \in R \ni g_{k,2} \geq |g_{0,k}(z_k)| \geq g_{k,1}$.

Assumption 4.5 For $k = 1, \ldots, N$ and $i = 1, \ldots, n$, disturbances ς_k are bounded, namely, $\exists\ D_{k,1} \ni D_{k,1} \geq |\varsigma_k|$.

Lemma 4.2 (Shen and Shi 2016; Shen et al. 2016, 2017, 2019) *For* $i = 1, \ldots, n$, $||\varpi_i|| \leq \frac{||e_i||}{\underline{\varsigma}(B+L)}$, *here matrix for matrix* $B + L$, $\underline{\varsigma}(B + L)$ *denotes its minimum singular value of, and e is the neighborhood synchronization error of the MASs, which will be defined later in (4.10).*

4.2.3 Backlash-Like Hysteresis

In this subsection, the mode of backlash-like hysteresis is firstly introduced. Then an assumption impacted the hysteresis is made, which is common and is widely used in the relevant results reported in literature.

In this work, the following backlash-like hysteresis H is investigated

$$w(t) = H(u(t)) \tag{4.3}$$

where $H(u) \in R$ and $u \in R$ are, respectively, the output and input of the hysteresis, which can be described

$$\frac{\partial w}{\partial t} = \Gamma \frac{\partial w}{\partial t} + \alpha \left| \frac{\partial u}{\partial t} \right| (-w + cu) \tag{4.4}$$

where $\alpha, c > 0$, and Γ are constants, respectively. Furthermore, $c > B$. Further, the solution of (4.4) can be described as follows:

$$w(t) = d_H(u) + cu(t) \tag{4.5}$$

where

$$\begin{aligned} d_H(u) =& e^{\alpha u sgn(u)} \int_{u_0}^{u} [-c + \Gamma] e^{\alpha \xi (sgn(u))} d\xi \\ &+ [-cu_0 + w_0] e^{-\alpha(-u_0+u)sgn(u)} \end{aligned}$$

Notice that, from the results in Shen and Shi (2016), it follows that the term $d_H(u)$ in the above Eq. (4.5) is bounded.

In order to distinguish the followers, subscript k is added to the symbols in (4.5). Then, substituting (4.5) into (4.1), it yields

$$\begin{cases} \dot{z}_{k,i} = z_{k,i+1}, \quad i = 1, \ldots, n-1 \\ \dot{z}_{k,n} = f_k(z_k) + g_{0,k}(z_k) c_k u_k + g_{0,k}(z_k) d_{H,k} + \varsigma_k(t) \\ \quad\;\; = f_k(z_k) + g_k(z_k) u_k + g_{0,k}(z_k) d_{H,k} + \varsigma_k(t) \\ y_k = z_{k,1} \end{cases} \tag{4.6}$$

where $g_k(z_k) = g_{0,k}(z_k) c_k$.

Assumption 4.6 For $k = 1, 2, \ldots, N$, $\varsigma_{H,k}$ and c_k are bounded, namely, $\exists$ unknown constants $D_{H,k} > 0 \in R$, $c_{\min,k} > 0 \in R$ and $c_{\max,k} > 0 \in R \ni c_k \in [c_{\min,k}, c_{\max,k}]$ and $D_{H,k} \geq |\varsigma_{H,k}|$.

4.2.4 Mathematical Description of Neural Networks

As stated in Chap. 2, as one of the universal approximations, neural networks (NNs) have been utilized widely in control design of dynamic systems, especially uncertain nonlinear systems (Hong et al. 2008). Radial basis function (RBF) NNs, as special NNs, will be approximated to the unknown continuous function derived from the closed-loop system stability analysis. The detailed description of RBF NNs can be found in Chap. 2 and omitted here.

In this chapter, NNs will be used to approximate the unknown functions in (4.6), i.e., $f_k(z_k))$ and $g_k(z_k))$. Over the region $\Omega_{z,k}$, by using the above neural networks, they are, respectively, represented as

$$g_k(z_k, W_{g,k}) = W_{g,k}^T \zeta_{g,k}, \quad f_k(z_k, W_{f,k}) = W_{f,k}^T \zeta_{f,k}. \tag{4.7}$$

Now, let us define the following sets and optimal weight vectors as

$$\Omega_{Wf,k} = \{W_{f,k} | \|W_{f,k}\| \leq M_{Wf,k}\},$$
$$\Omega_{Wg,k} = \{W_{g,k} | \|W_{g,k}\| \leq N_{Wg,k}\},$$

$$W^*_{f,k} = \arg \min_{W \in \Omega_{Wf,k}} [\sup_{z \in \Omega_{z,k}} |f_k(z_k) - f_k(z_k, W_{f,k})|], \text{ and}$$

$$W^*_{g,k} = \arg \min_{W \in \Omega_{Wg,k}} [\sup_{z \in \Omega_{z,k}} |g_k(z_k)| - g_k(z_k, W_{g,k})]$$

with design parameters $M_{Wg,k} > 0$ and $M_{Wf,k} > 0$ being real numbers. In addition, the optimal approximation error is represented as follows:

$$\kappa_k = [-g_k(z_k) + g_k(z_k, W^*_{g,k})] \times u_{f,k} - f_k(z_k) + f_k(z_k, W^*_{f,k}), \tag{4.8}$$

where

$$g_k(z_k, \hat{W}_{g,k}) = \hat{W}^T_{g,k}\zeta_{g,k}, \quad f_k(z_k, \hat{W}_{f,k}) = \hat{W}^T_{f,k}\zeta_{f,k}, \quad u_{f,k} = \frac{-f_k(z_k, \hat{W}_{f,k})}{g_k(z_k, \hat{W}_{g,k})},$$

and $\hat{W}_{g,k}$ and $\hat{W}_{f,k}$ denote respectively the estimates of $W^*_{g,k}$ and $W^*_{f,k}$.

Define

$$\varepsilon_{\kappa,k} = \max_{z_k \in \Omega_{z,k}, \hat{W}_{g,k} \in \Omega_{Wg,k}, \hat{W}_{f,k} \in \Omega_{Wf,k}} |\kappa_k| \tag{4.9}$$

Then, from the approximation theory of NNs, it follows that constant ε_κ is bounded although it is unknown.

Now, an assumption is given.

Assumption 4.7 For $k = 1, 2, \ldots, N$, the error $\varepsilon_{\kappa,k}$ is bounded on a compact set, namely, $|\varepsilon_{\kappa,k}| \leq \varepsilon^*$ with $\varepsilon^* > 0$ being an unknown constant.

Note that, it is for notational simplicity that $\bullet(\cdot)$ can be denoted by $\bullet$ in some cases. For example, for $f_k(z_k)$, its abbreviation is f_k.

Remark 4.1 In general, it is assumed that the region $\Omega_{z,k}$ is larger such that state x_k always be within it. As stated in the Introduction of this chapter, it is necessary for NNs' approximation theory that x always belongs to a compact set. However, how to determine the compact set is more difficult. Hence, in practical applications, design parameter $M_{x,k}$ is chosen to be a larger constant, which means that the region also is larger. Further, it can guarantee that x always be within the region, and ensure that NNs' approximation theory holds. Strictly speaking, this is incomplete. its determination method should be given, and the derivation and expression of the region also should be given. One of the main contributions in this work is to determine the compact set, which will be given in the following theorem.

4.3 Main Results

Recall the tracking error defined previously,

$$\varpi_{k,i} = z_{k,i} - z_{0,i}, \quad i = 1, 2, \ldots, n, \quad k = 1, 2, \ldots, N.$$

And set vectors ϖ_i, $\underline{x}_{0,i}$ and x_i as

$$z_i = [z_{1,i}, \ldots, z_{N,i}]^T, \quad \underline{z}_{0,i} = [z_{0,i}, \ldots, z_{0,i}]^T \in R^N \text{ and } \varpi_i = [\varpi_{1,i}, \ldots, \varpi_{N,i}]^T.$$

Further, we have

$$\varpi_i = z_i - \underline{z}_{0,i}$$

For follower k ($k = 1, 2, \ldots, N$), its ith neighborhood synchronization error is defined as

$$e_{k,i} = \sum\nolimits_{j \in \mathcal{N}_k} a_{j,k}(x_{j,i} - x_{k,i}) + b_k(x_{0,i} - x_{k,i}), \quad i = 1, \ldots, n. \tag{4.10}$$

For the sake of description, let's define the following symbols:

$$B = diag\{b_l\} \in R^{N \times N}, \quad e_i = [e_{1,i}, \ldots, e_{N,i}]^T,$$

$$\underline{f}_0 = [f_0, \ldots, f_0]^T \in R^N, \quad f = [f_1, \ldots, f_N]^T,$$

$$g = diag\{g_l\} \in R^{N \times N} \text{ with } g_l = g_{0,l} c_k, \quad l = 1, 2, \ldots, N,$$

$$h = [h_1, h_2, \ldots, h_N]^T \text{ with } h_l = g_{0,l} d_{H,l} + \varsigma_l, \quad l = 1, 2, \ldots, N.$$

Because of $L\underline{1} = 0$, it follows from (4.2), (4.5) and (4.10) that

$$\begin{cases} \dot{e}_i = e_{i+1} \\ \dot{e}_n = -(\mathscr{B} + \mathscr{L})(f + h + gu - \underline{f}_0) \end{cases} \tag{4.11}$$

where $i = 1, \ldots, n - 1$.

For follower k ($k = 1, \ldots, N$), define sliding mode error s_k as

$$s_k = c_1 e_{k,1} + c_2 e_{k,2} + \cdots + c_{n-1} e_{k,n-1} + e_{k,n} = \sum_{i=1}^{n-1} c_{k,i} e_{k,i} + e_{k,n} \tag{4.12}$$

where $c_{k,i} = C_{n-1}^{i-1} \alpha^{n-i}$, design parameter α_k is a positive real number, $C_{n-1}^{i-1} = \frac{(n-1)(n-2)\cdots(n-i+1)}{(i-1)(i-2)\cdots 1}$.

Now, let us define a vector

$$\bar{e}_k = [e_{k,1}, e_{k,2}, \ldots, e_{k,n}]^T \in R^N.$$

Lemma 4.3 *For follower k ($k = 1, 2, \ldots N$), considering the sliding mode error s_k represented in (4.12), it holds*
(1) if $s_k = 0$, then $\lim_{t\to\infty}\bar{e}_k(t) = 0$;
(2) if $\bar{e}_k(0) \in \Omega_{a_k}$ and $|s_k| \le a_k$, then for $\forall t \ge 0$, $\bar{e}_k(t) \in \Omega_{a_k}$;
(3) if $\bar{e}_k(0) \notin \Omega_{a_k}$ and $|s_k| \le a_k$, then $\exists T_k = (n-1)/\lambda_k$, $\ni \bar{e}_k(t) \in \Omega_{a_k}$ for $\forall t \ge T_k$, where $\Omega_{a_k} = \{\bar{e}_k(t) \mid |e_{k,m}| \le \lambda_k^{m-n} 2^{(m-1)} a_k$, $m = 1, \ldots, n$, design parameter a_k is a positive real number.

Just to reduce the complexity of the derivation, the design parameters are set as

$$\lambda_n = 1, \quad , c_{1,l} = c_{2,l} = \cdots = c_{N,l} = \lambda_l \quad l = 1, \ldots, n$$

then

$$s_k = \lambda_1 e_{k,1} + \cdots + \lambda_n e_{k,n}$$

For the overall system, define s as

$$s = [s_1, \ldots, s_N]^T$$

then

$$s = \lambda_1 e_1 + \cdots + \lambda_n e_n$$

$$\dot{s} = \lambda_1 \dot{e}_1 + \cdots + \lambda_n \dot{e}_n = \sum_{i=1}^{n-1} \lambda_i e_{i+1} + \dot{e}_n$$

Let $\gamma = \sum_{i=1}^{n-1} \lambda_i e_{i+1}$, it further follows

$$\dot{s} = -(\mathscr{B} + \mathscr{L})(h + f - \underline{f}_0 + gu) + \gamma \tag{4.13}$$

Define

$$\hat{f} = [\hat{f}_1, \hat{f}_2 \ldots, \hat{f}_N]^T \text{ with } \hat{f}_l = \hat{W}_{f,l}\zeta_{f,l}(z_l), \quad l = 1, \ldots, N,$$

$$f^* = [f_1^*, f_2^* \ldots, f_N^*]^T \text{ with } f_l^* = W_{f,l}^*\zeta_{f,l}(z_l), \quad l = 1, \ldots, N,$$

$$\hat{g} = [\hat{g}_1, \hat{g}_2 \ldots, \hat{g}_N]^T \text{ with } \hat{g}_l = \hat{W}_{g,l}\zeta_{g,l}(z_l), \quad l = 1, \ldots, N,$$

$$g^* = [g_1^*, g_2^* \ldots, g_N^*]^T \text{ with } g_l^* = W_{g,l}^*\zeta_{g,l}(z_l), \quad l = 1, \ldots, N.$$

From the definition of $u_{f,k}$ in the previous section, namely,

$$u_{f,k} = \frac{1}{g_k(z_k, \hat{W}_{g,k})}[-f_k(z_k, \hat{W}_{f,k})]$$

we have

$$f_k(x_k, \hat{W}_{f,k}) + u_{f,k} g_k(x_k, \hat{W}_{g,k}) = 0$$

Further, (4.13) is developed

$$\begin{aligned} \dot{s} &= (\mathscr{B} + \mathscr{L})(h - f + \underline{f}_0 - gu) + \gamma \\ &= (\mathscr{B} + \mathscr{L})[\tilde{f} + \tilde{g}u_f - gu_c - gu_s] \\ &\quad + (\mathscr{B} + \mathscr{L})(\underline{f}_0 + h) + \gamma \end{aligned} \tag{4.14}$$

where $\tilde{f} = \hat{f} - f$, $\tilde{g} = \hat{g} - g$, control input $u = u_f + u_s + u_c$, which is defined later in (4.17),

$$u_s = [u_{s,1}, \ldots, u_{s,N}]^T \in R^N, \;\; u_c = [u_{c,1}, \ldots, u_{c,N}]^T \in R^N, \;\; u_f = [u_{f,1}, \ldots, u_{f,N}]^T \in R^N.$$

Functions f_k and g_k are approximated by NNs, and (4.14) becomes

$$\begin{aligned} \dot{s} =& (\mathscr{B} + \mathscr{L})(h + \underline{f}_0) + \gamma \\ &+ (\mathscr{B} + \mathscr{L})[-\kappa + (\hat{f} - f^*) - gu_s - gu_c + (\hat{g} - g^*)u_f] \end{aligned} \tag{4.15}$$

where $\kappa = [\kappa_1, \kappa_2, \ldots, \kappa_N]^T$ with optimal approximation error κ_l being defined in (4.8), $l = 1, \ldots, N$.

Define

$$s_\tau = [s_{\tau,1}, s_{\tau,2}, \ldots, s_{\tau,N}]^T \text{ with } s_{\tau,k} = s_k - \tau sat(\frac{s_k}{\tau}) \tag{4.16}$$

where saturation function is abbreviated as symbol $sat(*)$, design parameter τ is a positive real number.

For follower k, $k = 1, 2, \ldots, N$, design the following control laws

$$u_k = u_{c,k} + u_{s,k} + u_{f,k}(z_k, \hat{W}_k), \tag{4.17}$$

where

$$u_{c,k} = \frac{1}{c_{\min,k} g_{k,1}} s_{\tau,k} + sat(\frac{s_k}{\tau})(b_k + d_k) \frac{g_{k,2} D_{k,h} + D_{k,1} + \hat{\varepsilon}_{\kappa.k}}{c_{\min,k} g_{k,1}}, \tag{4.18}$$

$$\begin{aligned} u_{s,k} =& sat(\frac{s_k}{\tau})(b_k + d_k) I_k(\bar{V}_k) \frac{1}{c_{\min,k} g_{k,1}} [F_k(z_k) \\ &+ |f_k(z_k, \hat{W}_{f,k})| + |g_k(x_k, \hat{W}_{g,k}) u_{f,k}| + |c_{\max,k} g_{k,2} u_{f,k}|], \end{aligned} \tag{4.19}$$

$$u_{f,k} = \frac{-f_k(z_k, \hat{W}_{f,k})}{g_k(z_k, \hat{W}_{g,k})}$$

where design parameter $\bar{V}_k$ is a positive real number, which is set as $\bar{V}_k = \frac{1}{2}\tau^2 > 0$; $\hat{\varepsilon}_{\kappa,k}$, $\hat{W}_{g,k}$ and $\hat{W}_{f,k}$ are the estimates of $\varepsilon_{\kappa,k}$, $W^*_{g,k}$, and $W^*_{f,k}$, respectively; and

$$I_k(\bar{V}_k) = \begin{cases} 1, \text{if } \; V_{s,k} = {s_{\tau,k}}^2/2 > \bar{V}_k; \\ 0, \text{if } \; V_{s,k} \le \bar{V}_k, \end{cases} \tag{4.20}$$

Or collectively

$$u = u_c + u_s + u_f(z, \hat{W}),$$

where $z = [z_1^T, \ldots, z_N^T]^T$,

$$\begin{aligned} u_c =& \frac{1}{c_{\min} G_1} s_\tau + \\ & (\mathscr{B} + \mathscr{D}) \frac{1}{c_{\min} G_1} sat(\frac{s}{\tau})(\hat{\varepsilon}_\nu + D_1 + G_2 D_2) \end{aligned} \tag{4.21}$$

$$\begin{aligned} u_s =& sat(\frac{s}{\tau})(\mathscr{B} + \mathscr{D}) I(\bar{V}) \frac{1}{c_{\min} G_1} [|\hat{f}| + F(z) + \\ & |\hat{g} u_f| + |c_{\max} G_2 u_f|] \end{aligned} \tag{4.22}$$

where $\frac{1}{c_{\min} G_1} = diag\{\frac{1}{c_{\min,1} g_{1,1}}, \ldots, \frac{1}{c_{\min,N} g_{N,1}}\}$, $sat(\frac{s}{\tau}) = diag\{sat(\frac{s_l}{\tau})\} \in R^{N \times N}$ with $l = 1, \ldots, N$, $s_\tau = [s_{\tau,1}, \ldots, s_{\tau,N}]^T \in R^N$, $D_1 + G_2 D_2 = [D_{1,1} + g_{1,2} D_{1,H}, \ldots, D_{N,1} + g_{N,2} D_{N,H}]^T$, $I(\bar{V}) = diag\{I_l(\bar{V}_l)\} \in R^{N \times N}$ with $l = 1, \ldots, N$, $|\hat{f}| = [|\hat{f}_1|, \ldots, |\hat{f}_N|]^T$, $F(z) = [F_1(z_1), \ldots, F_N(z_N)]^T$, $|\hat{g} u_f| = [|\hat{g}_1 u_{f,1}|, \ldots, |\hat{g}_N u_{f,N}|]^T \in R^N$, $|c_{\max} G_2 u_f| = [|c_{\max,1} g_{1,2} u_{f,1}|, \ldots, |c_{\max,N} g_{N,2} u_{f,N}|]^T \in R^N$, $\hat{\varepsilon}_\nu = [\hat{\varepsilon}_{\kappa,1}, \ldots, \hat{\varepsilon}_{\kappa,N}]^T$ with $\hat{\varepsilon}_{\kappa,l}$ being the estimate of $\varepsilon_{\kappa,l}$, $l = 1, \ldots N$.

For $\dot{\hat{W}}_{f,k}$, $\dot{\hat{W}}_{g,k}$ and $\dot{\hat{\varepsilon}}_{\kappa,k}$, $k = 1, \ldots, N$, design, respectively, the following adaptive laws,

$$\dot{\hat{W}}_{f,k} = \begin{cases} -\eta_{1,k} s_{\tau,k} \zeta_{f,k}, \\ \quad \text{if } ||\hat{W}_{f,k}|| < M_{Wf,k} \quad \text{or} \quad ||\hat{W}_{f,k}|| = M_{Wf,k} s_{\tau,k} \hat{W}_{f,k}^T \zeta_{f,k} \ge 0; \\ -\eta_{1,k} s_{\tau,k} \zeta_{f,k} + \eta_{1,k} s_{\tau,k} \frac{\hat{W}_{f,k} \hat{W}_{f,k}^T}{||\hat{W}_{f,k}||^2} \zeta_{f,k}, \\ \quad \text{if } ||\hat{W}_{f,k}|| = M_{Wf,k} s_{\tau,k} \hat{W}_{f,k}^T \zeta_{f,k} < 0, \end{cases}$$

$$g_k(z_k) = \hat{W}_{g,k}^T \zeta_{g,k}(z_k) \tag{4.23}$$

where $\hat{W}_{g,k} = [\hat{W}_{g,k,1}, \hat{W}_{g,k,2}, \ldots, \hat{W}_{g,k,N_{g,k}}]^T$, $N_{g,k}$ is the input dimension number of the NNs used to approximate $g_k(z_k)$. When $\hat{y}_{gl} = \varepsilon$,

$$\dot{\hat{W}}_{g,k,l} = \begin{cases} 0, & s_{\tau,k} u_{f,k} \zeta_{g,k,l} \geq 0 \\ -\eta_{2,k} s_{\tau,k} u_{f,k} \zeta_{g,k,l}, & s_{\tau,k} u_{fk,} \zeta_{gk,l} < 0; \end{cases}, \quad l = 1, \ldots, N_{g,k}, \tag{4.24}$$

otherwise

$$\dot{\hat{W}}_{g+,k} = \begin{cases} -\eta_{2,k} s_{\tau,k} u_{f,k} \zeta_{g+,k}, \ \text{if } ||\hat{W}_{g,k}|| < M_{Wg,k} \\ \quad \text{or } ||\hat{W}_{g,k}|| = M_{Wg,k} s_{\tau,k} u_{f,k} [\hat{W}_{g+,k}^T \zeta_{g+,k} + \hat{W}_{g\varepsilon1,k}^T \zeta_{g\varepsilon1,k}] \geq 0; \\ -\eta_{2,k} s_{\tau,k} u_{f,k} \zeta_{g+,k} + \eta_{2,k} s_{\tau,k} u_{f,k} \hat{W}_{g+,k} [\hat{W}_{g+,k}^T \zeta_{g+,k} \\ \quad + \hat{W}_{g\varepsilon1,k}^T \zeta_{g\varepsilon1,k}] / ||\hat{W}_{g+,k}||^2, \\ \quad \text{if } ||\hat{W}_{g,k}|| = M_{Wg,k} s_{\tau,k} u_{f,k} [\hat{W}_{g+,k}^T \zeta_{g+,k} + \hat{W}_{g\varepsilon1,k}^T \zeta_{g\varepsilon1,k}] < 0 \end{cases} \tag{4.25}$$

$$\dot{\hat{\varepsilon}}_{\kappa,k} = \eta_{3,k} |s_{\tau,k}|, \tag{4.26}$$

Or the adaptive laws are defined in the following compact form,

$$\dot{\hat{W}}_f = \begin{cases} -\eta_1 \zeta_f P(\mathscr{D} + \mathscr{B}) s_\tau, \\ \quad \text{if } ||\hat{W}_f|| < M_{Wf} \ \text{ or } ||\hat{W}_f|| = M_{Wf} \text{ and } s_\tau \hat{W}_f^T \zeta_f \geq 0; \\ -\eta_1 \zeta_f P(\mathscr{D} + \mathscr{B}) s_\tau + \eta_1 \zeta_f \dfrac{\hat{W}_f \hat{W}_f^T}{||\hat{W}_f||^2} s_\tau, \\ \quad \text{if } ||\hat{W}_f|| = M_{Wf} \text{ and } s_\tau \hat{W}_f^T \zeta_f < 0, \end{cases} \tag{4.27}$$

$$\dot{\hat{W}}_{g+} = \begin{cases} -\eta_2 \zeta_{g+} P(\mathscr{D} + \mathscr{B}) s_\tau, \ \text{if } ||\hat{W}_g|| < M_{Wg} \text{or} \\ \quad ||\hat{W}_g|| = M_{Wg} \text{ and } s_\tau u_f [\hat{W}_{g+} \zeta_{g+}^T + \hat{W}_{g\varepsilon1} \zeta_{g\varepsilon1}^T] \geq 0; \\ -\eta_2 \zeta_{g+} P(\mathscr{D} + \mathscr{B}) s_\tau + \\ \quad \eta_2 u_f \hat{W}_{g+} [\hat{W}_{g+} \zeta_{g+}^T + \hat{W}_{g\varepsilon1} \zeta_{g\varepsilon1}^T] / ||\hat{W}_{g+}||^2 s_\tau, \\ \quad \text{if } ||\hat{W}_g|| = M_{Wg} \text{ and} \\ \qquad s_\tau u_f [\hat{W}_{g+} \zeta_{g+}^T + \hat{W}_{g\varepsilon1} \zeta_{g\varepsilon1}^T] < 0 \end{cases} \tag{4.28}$$

$$\dot{\hat{\varepsilon}}_v = \eta_3 P(\mathscr{D} + \mathscr{B}) s_\tau \tag{4.29}$$

and the adaptive law of $\hat{\hat{M}}_{hf}$ is defined as

$$\dot{\hat{\hat{M}}}_{hf} = \eta_4 \text{sgn}(s_\tau^T P(\mathscr{B} + \mathscr{D})) P(\mathscr{B} + \mathscr{D}) s_\tau \tag{4.30}$$

where design parameter ε is a positive real number; $\eta_{l,k} > 0$, $l = 1, 2, 3$, denotes adaptive rates; $\zeta_{g+,k}$ and $\zeta_{g\varepsilon 1,k}$ are the member functions of NNs, which related, respectively, to $\hat{W}_{g+,k}$ and $\hat{W}_{g\varepsilon 1,k}$; $\hat{W}_{g+,k}$ denotes the collection of deleting the elements subject to (4.21) in $\hat{W}_{g,k,l}$.

In the following analysis, the boundedness of system states $x_k(t), k = 1, 2, \ldots, N$, first is considered. It is proven that system states $x_k(t)$ always belong to a compact set if the initial system states $x_k(0)$ is bounded and belong to the compact set. Based on it, the second result, i.e., the consensus control design, is investigated.

(1) Define

$$V_{1,k}(t) = \frac{1}{2}\hat{W}_{f,k}^T \hat{W}_{f,k}, \quad k = 1, 2, \ldots, N$$

Similar to Wang (1994), it is proved that, if initial value is bounded, i.e., $\hat{W}_{f,k}(0) \in \Omega_{Wf,k}$, we have

$$\hat{W}_{f,k}(t) \in \Omega_{Wf,k}, \quad k = 1, \ldots, N, \quad \forall t \geq 0$$

Similarly, for $k = 1, \ldots, N$, let us define

$$\begin{aligned} V_{2,k} &= \frac{1}{2}\hat{W}_{g,k}^T \hat{W}_{g,k} \\ &= [\hat{W}_{g+,k}^T \hat{W}_{g+,k} + \hat{W}_{g\varepsilon 1,k}^T \hat{W}_{g\varepsilon 1,k} \end{aligned}$$

Its time derivative can be described

$$\dot{V}_{2,k}(t) = \hat{W}_{g+,k}^T \dot{\hat{W}}_{g+,k} + \hat{W}_{g\varepsilon 1,k}^T \dot{\hat{W}}_{g\varepsilon 1,k}$$

Case I: If

$$||\hat{W}_{g,k}|| = M_{Wg,k} s_{\tau,k} u_{f,k} [\hat{W}_{g\varepsilon 1,k}^T \zeta_{g\varepsilon 1,k} + \hat{W}_{g+,k}^T \zeta_{g+,k}] \geq 0$$

then

$$\dot{V}_{2,k} = -\eta_{2,k} s_{\tau,k} u_{f,k} \hat{W}_{g+,k}^T \zeta_{g+,k} - \eta_{2,k} s_{\tau,k} u_{f,k} \hat{W}_{g\varepsilon 1,k}^T \zeta_{g\varepsilon 1,k} \leq 0$$

Case II: If

$$||\hat{W}_{g,k}|| = M_{Wg,k} s_{\tau,k} u_{f,k} [\hat{W}_{g+,k}^T \zeta_{g+,k} + \hat{W}_{g\varepsilon 1,k}^T \zeta_{g\varepsilon 1,k}] < 0$$

then we have

$$\dot{V}_{2,k} = 0$$

From the analysis in **Cases I** and **II**, it can be obtained that, if $\hat{W}_{g,k}(0) \in \Omega_{Wg,k}$, then $||\hat{W}_{g,k}|| \leq M_{Wg,k}$, $\forall t \geq 0$.

Therefore, for $k = 1, \ldots, N$, $\hat{W}_{f,k}(0) \in \Omega_{Wf,k}$ and $\hat{W}_{g,k}(0) \in \Omega_{Wg,k}$, it can be obtained that

$$\text{for } \forall t \geq 0, \quad \hat{W}_{f,k}(t) \in \Omega_{Wf,k} \text{ and } \hat{W}_{g,k}(t) \in \Omega_{Wg,k}.$$

(2) Beginning with the following analysis, $\dot{s}_{\tau,k}$ is first considered.
From (4.16), it follows

$$s_{\tau,k} = s_k - \tau sat(\frac{s_k}{\tau})$$

After analyzing $s_{\tau,k}$, we can obtain
(1) $s_{\tau,k} = 0$ if $|s_k| < \tau$;
(2) $\frac{ds^2_{\tau,k}}{dt} = 0$ and $s_{\tau,k} = 0$ if $|s_k| = \tau$;
(3) $\dot{s}_{\tau,k} = \dot{s}_k$ and $|s_{\tau,k}| = |s_k| - \tau$.
Define

$$V_{s,\tau} = \frac{1}{2} s_\tau^T P s_\tau$$

where matrix P is positively definite symmetric, which is defined in Lemma 4.1.
From the analysis in (1), (2), and (3), we know, (a) $\dot{V}_{\tau,k} = 0$ if $|s_{\tau,k}| \leq \tau$. On the other hand, if $s_{s,\tau} > \tau$, it follows from (4.14)that

$$\begin{aligned}
\dot{V}_{s,\tau}(t) &= s_\tau^T P \dot{s}_\tau \\
&= s_\tau^T P[\gamma + (\mathscr{B} + \mathscr{L})(\underline{f}_0) + h - gu - f] \\
&= s_\tau^T P\{\gamma + (\mathscr{B} + \mathscr{L})[(\hat{g} - g)u_f + (\hat{f} - f) - gu_s \\
&\quad - gu_c]\} + s_\tau^T P[(\mathscr{B} + \mathscr{L})(\underline{f}_0) + h] \\
&= s_\tau^T P\{\gamma + (\mathscr{B} + \mathscr{D})[(\hat{f} - f) + (\hat{g} - g)u_f - gu_c - gu_s]\} \\
&\quad + s_\tau^T P[(\mathscr{B} + \mathscr{D})(h + \underline{f}_0)] \\
&\quad + s_\tau^T P\{-A[(\hat{g} - g)u_f + (\hat{f} - f) - gu_c - gu_s]\} \\
&\quad + s_\tau^T P[-A(\underline{f}_0 + h)]
\end{aligned} \tag{4.31}$$

Note that, $\mathscr{L} = \mathscr{D} - \mathscr{A}$.
Substituting (4.17), (4.18)–(4.20) into (4.27), and from

$$|s_{\tau,k}| = s_{\tau,k} sat(\frac{s_k}{\tau}), \quad \text{and} \quad s_{\tau,k}\dot{s}_{\tau,k} = s_{\tau,k}\dot{s}_k$$

one has

$$\begin{aligned}\dot{V}_{s,\tau}(t) \leq & -s_\tau^T P(\mathscr{B}+\mathscr{L})s_\tau + |s_\tau|\Delta \\ & -\frac{s_\tau^T g W}{c_{\min} G_1} sat(\frac{s}{\tau}) - \frac{s_\tau^T g \hat{\varepsilon}_v}{c_{\min} G_1} sat(\frac{s}{\tau}) \\ & -\frac{s_\tau^T g(D_1+G_2 D_2)}{c_{\min} G_1} sat(\frac{s}{\tau}) + |s_\tau|(D_1+G_2D_2) \\ \leq & \, 0 \end{aligned} \tag{4.32}$$

where $|s_\tau| = [|s_{\tau,1}|, \ldots, |s_{\tau,N}|]$, $\Delta = [\Delta_1, \ldots, \Delta_N]^T$, $\Delta_m = |c_{\max,m} g_{m,2} u_{f,m}| + F_m(z_m) + |g_m(z_m, \hat{W}_{g,m}) u_{f,m}| + |f_m(z_m, \hat{W}_{f,m})|$, $m = 1, \ldots, N$.

Therefore, the following results can be obtained that, if $V_{s,\tau} > \bar{V}$, then $\dot{V}_{s,\tau} \leq 0$. In addition, because $V_{s,\tau}$ is continuous, the designers can obtain the value of $V_{s,\tau}$ if sampling period is set to be small sufficiently. Further, determines whether the function $V_{s,\tau}$ is less than a constant $2\bar{V}$, which further determines whether supervisory control u_s in (4.22) is used. In a word, under the supervisory control, we obtain

$$\text{for } \forall t \geq 0, \ \ ||s|| \leq 2\sqrt{\bar{V}}.$$

From Lemma 4.3, we know, for $k = 1, 2, \ldots N$ and $m = 1, \ldots, n$, $|e_{k,m}| \leq \lambda_k^{m-n} 2^{(m-1)} a_k$, $\forall t \geq 0$, then $e_{k,m} \in L_\infty$. We can obtain that

$$||z_k|| \leq M_{z,k} + \sum_{i=1}^{n} 2^{(i-1)} \lambda_k^{i-n} a_k, \tag{4.33}$$

Obviously, under the bounded initial condition, states z_k always belong to a compact set, namely, if $z_k(0) \in \Omega_{z,k}$, then

$$z_k(t) \in \Omega_{z,k}, \ \ \forall t > 0,$$

where

$$\Omega_{z,k} = \left\{ z_k(t) \mid ||z_k|| \leq M_{z,k} + \sum_{i=1}^{n} 2^{(i-1)} \lambda_k^{i-n} a_k \right\}.$$

Thus, the first result of this chapter has been proven.

Now, let us prove the second result in this chapter.

Define

$$V(t) = V_{s,\tau} + \frac{1}{2\eta_1} tr\{\tau_f^T \tau_f\} + \frac{1}{2\eta_2} tr\{\tau_g^T \tau_g\} + \frac{1}{2\eta_3} \tilde{\varepsilon}_\kappa^T \tilde{\varepsilon}_\kappa + \frac{1}{2\eta_4} \tilde{\bar{M}}_{hf}^T \tilde{\bar{M}}_{hf} \tag{4.34}$$

where $\tau_f = \hat{W}_f - W_f^*$, $\tau_g = \hat{W}_g - W_g^*$, $\tilde{\varepsilon}_\kappa = \hat{\varepsilon}_\kappa - \varepsilon_\kappa$, $\tilde{\bar{M}}_{hf} = \bar{M}_{hf} - \hat{\bar{M}}_{hf}$, $\hat{\bar{M}}_{hf} = [\hat{M}_{hf,1}, \ldots, \hat{M}_{hf,N}]^T$, $\hat{M}_{hf,k}$ is the estimate of $M_{hf,k} = g_{k,2} D_{k,H} + D_{k,1} + M_{f0}$, $k = 1, \ldots, N$, $\bar{M}_{hf} = [M_{hf,1}, \ldots, M_{hf,N}]^T$.

Further,

$$\dot{V}(t) = \dot{V}_{s,\tau} + \frac{1}{\eta_1} tr\{\tau_f^T \dot{\hat{W}}_f\} + \frac{1}{\eta_2} tr\{\tau_g^T \dot{\hat{W}}_g\} + \frac{1}{\eta_3} \tilde{\varepsilon}_\kappa^T \dot{\hat{\varepsilon}}_\kappa \frac{1}{\eta_4} \tilde{\bar{M}}_{hf}^T \dot{\hat{\bar{M}}}_{hf} \tag{4.35}$$

From the definitions of s and s_τ, it follows that

$$\begin{cases} (i)\ \dot{s}_\tau = \dot{s} \text{ if } |s| > \tau; \\ (ii)\ \dfrac{ds_\tau^2}{dt} = 0 \text{ and } s_\tau = 0 \text{ if } |s| = \tau. \end{cases}$$

From (4.20), the following result is obtained

$$I_k(V_k)) = 0 \text{ if } V_k \leq \bar{V}_k.$$

Further, from the definition of u_s in (4.22), we have

$$u_s = \underline{0} \text{ with } \underline{0} = [0, \ldots, 0] \in R^N.$$

Then, substituting (4.19) and (4.27) into (4.31), it yields

$$\begin{aligned} \dot{V} = & s_\tau^T P(\mathscr{D} + \mathscr{B})\tau_f^T \zeta_f - s_\tau^T P(\mathscr{L} + \mathscr{B}) s_\tau + \\ & s_\tau^T P(\mathscr{D} + \mathscr{B}) u_f \tau_g^T \zeta_g + s_\tau^T P(\mathscr{D} + \mathscr{B}) \tilde{\varepsilon}_\kappa + \frac{1}{\eta_3} \tilde{\varepsilon}_\kappa^T \dot{\hat{\varepsilon}}_\kappa + \\ & s_\tau^T P(\mathscr{D} + \mathscr{B}) sgn(s_\tau^T P(\mathscr{D} + \mathscr{B})) \tilde{\bar{M}}_{hf} - s_\tau^T P \mathscr{A} \tau_f^T \zeta_f - \\ & s_\tau^T P \mathscr{A} u_f \tau_g^T \zeta_g - s_\tau^T P \mathscr{A} \tilde{\varepsilon}_\kappa + s_\tau^T P[-\mathscr{A}(\underline{f}_0 + h)] - \\ & s_\tau^T P \mathscr{A} sgn(s_\tau^T P(\mathscr{D} + \mathscr{B})) \hat{\bar{M}}_{hf} + \frac{1}{\eta_4} \tilde{\bar{M}}_{hf}^T \dot{\hat{\bar{M}}}_{hf} + s_\tau^T P \mathscr{A} (\mathscr{D} + \mathscr{B}) \gamma + \\ & \frac{1}{\eta_1} tr\{\tau_f^T \dot{\hat{W}}_f\} + \frac{1}{\eta_2} tr\{\tau_g^T \dot{\hat{W}}_g\} \end{aligned} \tag{4.36}$$

Since

$$- s_\tau^T P \mathscr{A} \tau_f^T \zeta_f \leq \bar{\varsigma}(\mathscr{A}) \bar{\varsigma}(P) \times \left(r s_\tau^T s_\tau + \frac{\zeta_f^T \tau_f \tau_f^T \zeta_f}{4r} \right),$$

$$-s_\tau^T P \mathscr{A} u_f \tau_g^T \zeta_g \leq \bar{\varsigma}(\mathscr{A}) \bar{\varsigma}(P) \times \left(r s_\tau^T s_\tau + \frac{\zeta_g^T \tau_g u_f^T u_f \tau_g^T \zeta_g}{4r} \right),$$

$$- s_\tau^T P \mathscr{A} \tilde{\varepsilon}_\kappa \leq \bar{\varsigma}(\mathscr{A}) \bar{\varsigma}(P) \times \left(r s_\tau^T s_\tau + + \frac{\tilde{\varepsilon}_\kappa^T \tilde{\varepsilon}_\kappa}{4r} \right)$$

$$s_\tau^T P[-\mathscr{A}(\underline{f}_0 + h)] \le \bar{\varsigma}(\mathscr{A})\bar{\varsigma}(P) \times \left(rs_\tau^T s_\tau + \frac{\bar{M}_{hf}^T \bar{M}_{hf}}{4r}\right),$$

$$-s_\tau^T P\mathscr{A} sgn(s_\tau^T P(\mathscr{D} + \mathscr{B}))\hat{\bar{M}}_{hf} \le \bar{\varsigma}(\mathscr{A})\bar{\varsigma}(P) \times \left(rs_\tau^T s_\tau + \frac{\hat{\bar{M}}_{hf}^T \hat{\bar{M}}_{hf}}{4r}\right),$$

$$s_\tau^T P\mathscr{A}(\mathscr{B} + \mathscr{D})\gamma \le \bar{\varsigma}(\mathscr{A})\bar{\varsigma}(P) \times \left(rs_\tau^T s_\tau + \frac{\gamma^T \gamma}{4r}\right),$$

where design parameter r is a positive real number, one has

$$\begin{aligned}
\dot{V}(t) \le & s_\tau^T P(\mathscr{D} + \mathscr{B})\tau_f^T \zeta_f - s_\tau^T P(\mathscr{L} + \mathscr{B})s_\tau + \\
& s_\tau^T P(\mathscr{D} + \mathscr{B})u_f \tau_g^T \zeta_g + s_\tau^T P(\mathscr{D} + \mathscr{B})\tilde{\varepsilon}_\kappa + \frac{1}{\eta_3}\tilde{\varepsilon}_\kappa^T \dot{\hat{\varepsilon}}_\kappa + \\
& s_\tau^T P(\mathscr{D} + \mathscr{B}) sgn(s_\tau^T P(\mathscr{D} + \mathscr{B}))\tilde{\bar{M}}_{hf} + \frac{1}{\eta_4}\tilde{\bar{M}}_{hf}^T \dot{\hat{\bar{M}}}_{hf} + \\
& 6\bar{\varsigma}(\mathscr{A})\bar{\varsigma}(P)\left[rs_\tau^T s_\tau + \frac{1}{4r}\left(\zeta_f^T \tau_f \tau_f^T \zeta_f + \zeta_g^T \tau_g u_f^T u_f \tau_g^T \zeta_g + \bar{M}_{hf}^T \bar{M}_{hf}\right)\right] + \\
& \bar{\varsigma}(\mathscr{A})\bar{\varsigma}(P)\left(\frac{1}{4r}\gamma^T \gamma + \tilde{\varepsilon}_\kappa^T \tilde{\varepsilon}_\kappa + \hat{\bar{M}}_{hf}^T \hat{\bar{M}}_{hf}\right) + \frac{tr\{\tau_f^T \dot{\hat{W}}_f\}}{\eta_1} + \frac{tr\{\tau_g^T \dot{\hat{W}}_g\}}{\eta_2}
\end{aligned} \tag{4.37}$$

From the definition of adaptive law defined in (4.30), it is easy to find, $\hat{\bar{M}}_{hf}$ is bounded. And Assumptions 4.2, 4.5 and 4.6 ensure the boundedness of h_k and f_0. Hence, $\tilde{\bar{M}}_{hf}$ is also bounded.

Because of the boundedness of $\tau_f^T \zeta_f, u_f \tau_g^T \zeta_g, \bar{M}_{hf}, \tilde{\varepsilon}_\kappa$ and $\hat{\bar{M}}_{hf}$, we can obtain the results: if design parameter $r > 0 \in R$ is chosen appropriately, then the following inequality holds

$$\frac{\bar{\varsigma}(\mathscr{A})\bar{\varsigma}(P)}{4r}(\zeta_g^T \tau_g u_f^T u_f \tau_g^T \zeta_g + \zeta_f^T \tau_f \tau_f^T \zeta_f + \tilde{\varepsilon}_\kappa^T \tilde{\varepsilon}_\kappa + \bar{M}_{hf}^T \bar{M}_{hf} + \hat{\bar{M}}_{hf}^T \hat{\bar{M}}_{hf}) \le \mu_s$$

where design parameter$\mu_s > 0 \in R$. Because

$$\gamma_k^2 = \sum\nolimits_{i=1}^{n-1} \lambda_i^2 e_{i+1}^2 \le \frac{\bar{\lambda}^2}{\underline{\lambda}^2} s_k^2$$

where $\bar{\lambda} = \max\{\lambda_1, \dots, \lambda_n\}$, $\underline{\lambda} = \min\{\lambda_1, \dots, \lambda_n\}$, one has

$$\gamma^T \gamma = \sum\nolimits_{k=1}^{N} \gamma_k^2 \le \sum\nolimits_{k=1}^{N} \frac{\bar{\lambda}^2}{\underline{\lambda}^2} s_k^2 = \frac{\bar{\lambda}^2}{\underline{\lambda}^2} s^T s$$

Further, (4.37) can be developed as

$$\begin{aligned}\dot{V} \leq & s_\tau^T P(\mathscr{D}+\mathscr{B})\tau_f^T \zeta_f - s_\tau^T P(\mathscr{B}+\mathscr{L})s_\tau + \\ & s_\tau^T P(\mathscr{D}+\mathscr{B})u_f \tau_g^T \zeta_g + s_\tau^T P(\mathscr{D}+\mathscr{B})\tilde{\varepsilon}_\kappa + \\ & (6r + \frac{\bar{\lambda}^2}{4r\underline{\lambda}^2})\bar{\varsigma}(\mathscr{A})\varsigma(P) r s_\tau^T s_\tau + \mu_s + \\ & s_\tau^T P(\mathscr{D}+\mathscr{B}) sgn(s_\tau^T P(B+D))\tilde{\bar{M}}_{hf} + \\ & \frac{1}{\eta_1} tr\{\tau_f^T \dot{\hat{W}}_f\} + \frac{1}{\eta_2} tr\{\tau_g^T \dot{\hat{W}}_g\} + \frac{1}{\eta_3}\tilde{\varepsilon}_\kappa^T \dot{\hat{\varepsilon}}_\kappa + \frac{1}{\eta_4}\tilde{\bar{M}}_{hf}^T \dot{\hat{\bar{M}}}_{hf}\end{aligned} \tag{4.38}$$

Substituting the adaptive laws (4.27)–(4.30) into (4.38), it yields

$$\begin{aligned}\dot{V}(t) \leq & - s_\tau^T P(\mathscr{B}+\mathscr{L})s_\tau + \left(6r + \frac{\bar{\lambda}^2}{4r\underline{\lambda}^2}\right) \\ & \times \bar{\varsigma}(\mathscr{A})\bar{\varsigma}(P) r s_\tau^T s_\tau + \mu_s - I_1 \eta_1 s_\tau^T \tau_f^T \frac{\hat{W}_f \hat{W}_f^T}{||\hat{W}_f||^2}\zeta_f \\ & + I_2 \eta_2 s_\tau^T \tau_{g+}^T \hat{W}_{g+}[\hat{W}_{g+}^T \zeta_{g+} + \hat{W}_{g\varepsilon 1}^T \zeta_{g\varepsilon 1}]/||\hat{W}_{g+}||^2\end{aligned} \tag{4.39}$$

where column vector τ_{g+} is composed of the elements of τ_g related to $\hat{W}_{g+}$; I_1 and I_2 are defined as:

$$I_1 = \begin{cases} 0, & \text{if condition 1 in (4.27) holds} \\ 1, & \text{if condition 2 in (4.27) holds} \end{cases},$$

$$I_2 = \begin{cases} 0, & \text{if condition 1 in (4.28) holds} \\ 1, & \text{if condition 2 in (4.28) holds.} \end{cases}$$

From the definition of (4.27), we know, if condition 2 in (4.27) is satisfied,

$$||\hat{W}_f|| = M_{Wf} \text{ and } s_\tau^T \hat{W}_f^T \zeta_f < 0.$$

Further, one have

$$\tau_f^T \hat{W}_f = [||\hat{W}_f||^2 - ||W_f^*||^2 + ||W_f^* - \hat{W}_f||^2]/2 \geq 0;$$

Similarly, if condition 2 in (4.28) is satisfied, then

$$||\hat{W}_g|| = M_{Wg} \text{ and } s_\tau^T u_f [\hat{W}_{g+}^T \zeta_{g+} + \hat{W}_{g\varepsilon 1}^T \zeta_{g\varepsilon 1}] < 0.$$

Further, one has

$$\frac{1}{2}\tau_{g+}^T\hat{W}_{g+} = [||\hat{W}_{g+}||^2 - ||W_{g+}^*||^2 + ||W_{g+}^* - \hat{W}_{g+}||^2] \geq 0.$$

Hence, one has

$$\dot{V}(t) \leq -\lambda_0 V + \mu_0 \tag{4.40}$$

where $\mu_0 = \mu_s$ and

$$\lambda_0 = \min\left\{\left(\frac{1}{2\bar{\varsigma}(P)}\underline{\varsigma}(Q) - \bar{\varsigma}(P)\bar{\varsigma}(A) \times (6r + \frac{1}{4r}\frac{\bar{\lambda}^2}{\underline{\lambda}^2})\right)\right\}.$$

Further, it follows

$$\begin{aligned} V(t) &\leq \frac{\mu_0}{\lambda_0} + (V(t_0) - \frac{\mu_0}{\lambda_0})e^{-\lambda_0(t-t_0)} \\ &\leq \frac{\mu_0}{\lambda_0} + V(t_0) = \varepsilon_V \end{aligned}$$

Since $2V(t) \geq 2V_{s,\tau}(t) \geq \underline{\varsigma}(P)||s(t)||^2$, one has

$$||s(t)|| \leq \sqrt{2\varepsilon_V/\underline{\varsigma}(P)}$$

From the definition of $V(t)$, it can be further obtained

$$||\tilde{W}_f(t)|| \leq \sqrt{2\eta_1\varepsilon_V/\underline{\varsigma}(P)},$$

$$||\tilde{W}_g(t)|| \leq \sqrt{2\eta_2\varepsilon_V/\underline{\varsigma}(P)},$$

$$||\tilde{\tilde{M}}_{hf}|| \leq \sqrt{2\eta_4\varepsilon_V/\underline{\varsigma}(P)},$$

$$||\ddot{\varepsilon}_\kappa|| \leq \sqrt{2\eta_3\varepsilon_V/\underline{\varsigma}(P)}.$$

Because $||s(t)|| \leq \sqrt{2\varepsilon_V/\underline{\varsigma}(P)}$ and $s = [s_1, \ldots, s_N]^T$, one has

$$|s_k(t)| \leq \sqrt{\frac{1}{\underline{\varsigma}(P)}2\varepsilon_V}$$

According to Lemma 4.3, it yields, for $k = 1, \ldots, N$,

$$|e_{k,m}(t)| \leq 2^{(m-1)}\lambda_k^{(m-n)}\sqrt{\underline{\varsigma}(P) \times 2\varepsilon_V}$$

where $m = 1, \ldots, n$. Because $e_m = [e_{1,m}, \ldots, e_{N,m}]^T \in R^N$, it follows

$$||e_m(t)|| \leq \sqrt{N \times \left(2^{(m-1)}\lambda_k^{(m-n)}\sqrt{\frac{1}{\underline{\varsigma}(P)2\varepsilon_v}}\right)^2}.$$

Finally, from Lemma 4.2, it can be obtained that, for $i = 1, \ldots, n$, $k = 1, \ldots, N$,

$$||\varpi_i(t)|| \leq \frac{1}{\underline{\varsigma}(L+B)} \times \sqrt{N \times \left(2^{(m-1)}\lambda_k^{(m-n)}\sqrt{\frac{1}{\underline{\varsigma}(P)2\varepsilon_v}}\right)^2}$$

which means that each follower synchronizes to the leader, and all signals including tracking errors converge to a neighborhood of the origin.

A theorem summarizes the above analysis.

Theorem 4.1 *Considering the leader-following system (4.1) and (4.6), if Assumptions 4.1–4.7 hold, and control laws (4.17)–(4.20) and adaptive laws (4.23)–(4.26) with (4.30) or (4.27)–(4.29) with (4.30) are applied, then*

(1) under bounded initial condition, each follower's states are bounded; and

(2) each follower synchronizes to the leader, and the signals in the closed-loop system converge to an adjustable neighborhood of the origin.

Proof Based on the analysis mentioned above, it can be obtained the conclusion. Hence, the proof is not given again here.

4.4 Example

In this work, let us consider a leader-following system, which has four inverted pendulums. The inverted pendulums are taken as followers. A virtual leader will be described later. The inverted pendulums and the virtual leader are described as a directed graph. Figure 4.1 shows system topology, where the weights are on the edges. It is assumed that each follower has the same form of (4.1), namely,

$$\begin{cases} \dot{z}_{k,i} = z_{k,i+1}, \quad i = 1, 2, \ldots, n-1 \\ \dot{z}_{k,n} = f_k(z_k) + g_{0,k}(z_k)w_k(u_k) + \varsigma_k(t) \\ y_k = z_{k,1} \end{cases}$$

where

$$f_k(z_k) = \frac{g\sin(z_{k,1}) - 0.5m_kL_kz_{k,2}^2\sin(2z_{k,1})(M_k+m_k)^{-1}}{L_k(\frac{4}{3} - (M_k+m_k)^{-1}m_k\cos^2(z_{k,1}))}$$

$$g_{0,k}(z_k) = \frac{(M_k + m_k)^{-1}\cos(z_{k,1})}{L_k(\frac{4}{3} - (M_k + m_k)^{-1} m_k \cos^2(z_{k,1}))}$$

where $z_{k,1}$ denotes angular position and $z_{k,2}$ denotes speed of the pole, m_k and M_k denotes respectively pole and cart masses. L_k also denotes half-length of the pole. In this simulation, the parameters are set as

$$\begin{aligned}
M_0 &= 0.75\,\text{kg},\ m_1 = 0.07\,\text{kg},\ L_1 = 0.45\,\text{m}\\
M_1 &= 0.8\,\text{kg},\ m_1 = 0.08\,\text{kg},\ L_1 = 0.40\,\text{m}\\
M_2 &= 0.9\,\text{kg},\ m_2 = 0.09\,\text{kg},\ L_2 = 0.45\,\text{m}\\
M_3 &= 1.0\,\text{kg},\ m_3 = 0.10\,\text{kg},\ L_3 = 0.50\,\text{m}\\
M_4 &= 1.1\,\text{kg},\ m_4 = 0.11\,\text{kg},\ L_4 = 0.55\,\text{m}
\end{aligned}$$

and ς_k is bounded, which is taken randomly by $|\varsigma_k| \le 1$, $k = 1, 2, 3, 4$. The leader is described as

$$\begin{cases} \dot{z}_{0,1} = z_{0,2} \\ \dot{z}_{0,2} = cos(t) \end{cases} \tag{4.41}$$

In addition, the parameters in the hysteresis (4.3) are chosen as

$$\begin{aligned}
a_1 &= 1,\ c_1 = 3.1635,\ B_1 = 0.34\\
a_2 &= 1,\ c_2 = 3.1636,\ B_2 = 0.35\\
a_3 &= 1,\ c_3 = 3.1634,\ B_3 = 0.36\\
a_4 &= 1,\ c_4 = 3.1632,\ B_4 = 0.32
\end{aligned}$$

The initial states are set as: $[x_0, v_0]^T = [0.1, -0.1]^T$, $[x_1, v_1]^T = [-0.1, -0.15]^T$, $[x_2, v_2]^T = [0.15, 0.15]^T$, $[x_3, v_3]^T = [-0.15, -0.1]^T$ and $[x_4, v_4]^T = [0.2, 0.1]^T$. For NNs, $\hat{W} \in R^{10}$ are randomly chosen in (0, 1], $c_{min} = 1$ and $c_{max} = 3$, and sample time is set to be 0.08s.

Partial results are given by Figs. 4.2, 4.3, 4.4 and 4.5. From Fig. 4.2, it is obvious that the states of the leader are bounded. From Figs. 4.3, 4.4 and 4.5, we also can find, under the control and adaptive laws (4.17–4.30), the followers can synchronize to the leader, which is shown in Fig. 4.3. From Fig. 4.4, it is easy to find that the synchronization errors converge asymptotically to an adjustable neighborhood of the origin. Figure 4.5 shows that the control inputs of the followers are bounded. Note that, Fig. 5.3 also shows that the states of the followers are bounded under supervisory control.

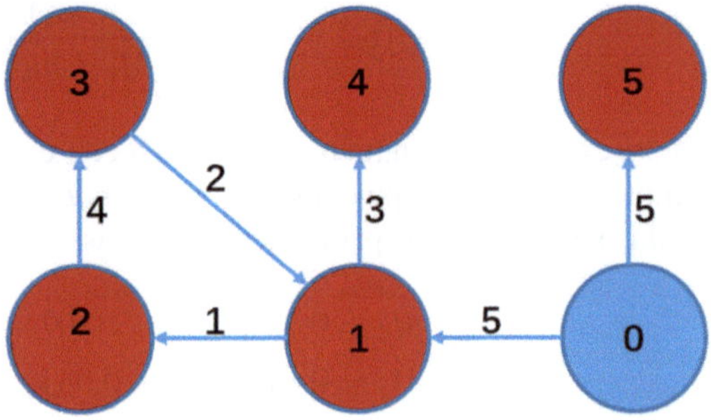

Fig. 4.1 Topology $\overline{G}$

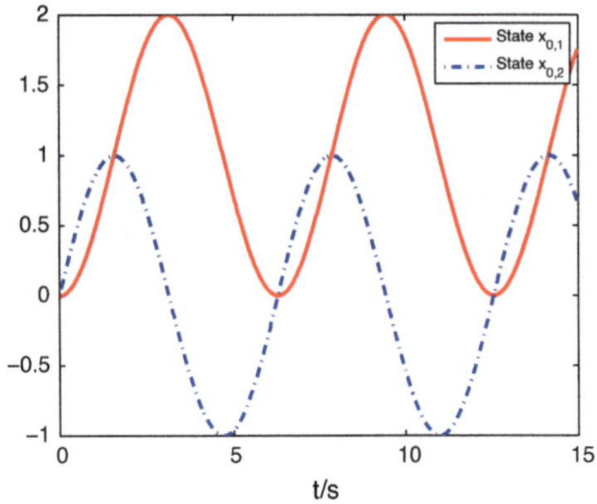

Fig. 4.2 The states of the leader

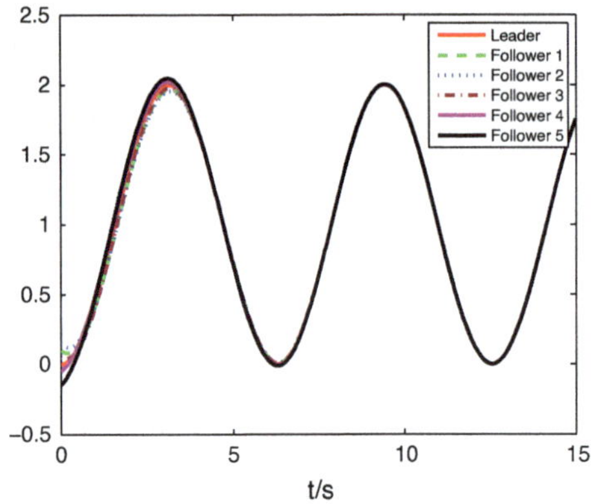

Fig. 4.3 The outputs of the leader and followers

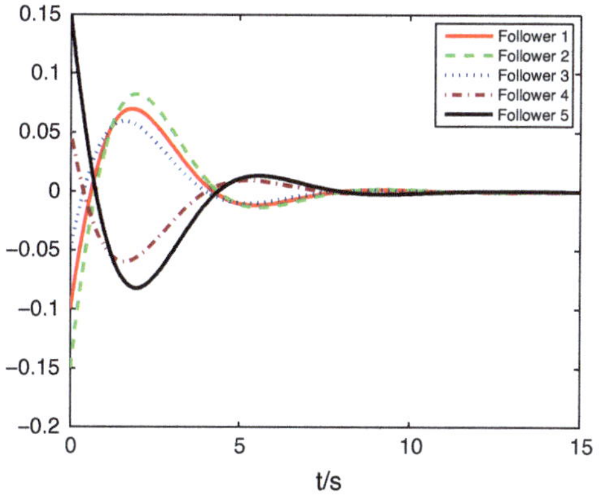

Fig. 4.4 The synchronization errors of the followers

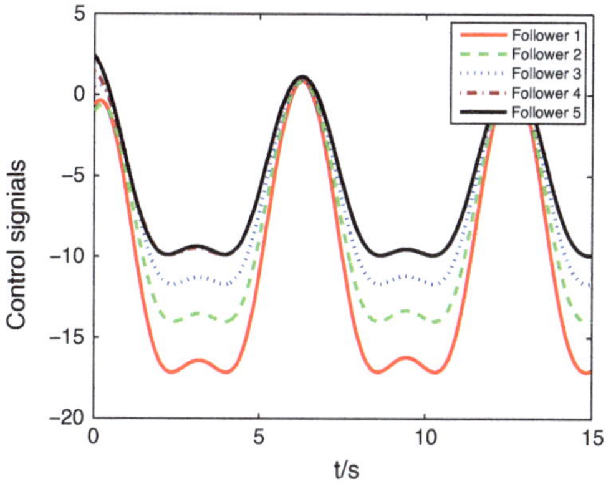

Fig. 4.5 The control inputs of the followers

4.5 Conclusion

This chapter investigates the consensus problem of uncertain leader-following systems subjected to backlash-like hysteresis, which is modeled as a directed graph. By using adaptive control, neural networks, and other methods, an adaptive control strategy with supervisory is constructed to guarantee that the consensus objective is achieved.

References

Chwastek K, Gozdur R (2019) Towards a unified approach to hysteresis and micromagnetics modeling: a dynamic extension to the Harrison model. Physica B: Condensed Matter 572(1):242–246

Defoort M, Floquet T, Kokosy A, Perruquetti W (2008) Sliding-mode formation control for cooperative autonomous mobile robots. IEEE Trans Ind Electron 55(11):3944–3953

Elbohy H, El-Mahalawy H, El-Ghamaz NA, Zidan H (2019) Hysteresis analysis in dye-sensitized solar cell based on different metal alkali cations in the electrolyte. Electrochimica Acta 319(1):110–117

Hong Y, Chen G, Bushnell L (2008) Distributed observers design for leader-following control of multi-agent networks. Automatica 44(3):846–850

Ikhouane F (2020) Theory of continuous rate-dependent hysteresis. Commun Nonlinear Sci Numer Simul (80). https://doi.org/10.1016/j.cnsns.2019.104970

Park K, Kim S, Lee H, Park I, Jung K (2019) Low-hysteresis and low-interference soft tactile sensor using a conductive coated porous elastomer and a structure for interference reduction. Sens Actuators A: Phys 295(15):541–550

Shen QK, Shi P (2016) Output consensus control of multiagent systems with unknown nonlinear dead zone. IEEE Trans Syst Man Cybern: Syst 46(10):1329–1337

Shen QK, Jiang B, Cocquempot V (2013) Fuzzy logic system-based adaptive fault tolerant control for near space vehicle attitude dynamics with actuator faults. IEEE Trans Fuzzy Syst 21(2):289–300

Shen QK, Jiang B, Shi P (2014) Cooperative adaptive fuzzy tracking control for networked unknown nonlinear multi-agent systems with time-varying actuator faults. IEEE Trans Fuzzy Syst 22(3):494–504

Shen QK, Jiang B, Shi P (2014) Fault diagnosis for T-S fuzzy systems with sensor faults and system performance analysis. IEEE Trans Fuzzy Syst 22(2):274–285

Shen QK, Shi P, Shi Y (2016) Distributed adaptive fuzzy control for nonlinear multiagent systems via Sliding mode observers. IEEE Trans Cybern 46(12):3086–3097

Shen QK, Shi P, Shi Y, Zhang J (2017) Adaptive output consensus with saturation and dead-zone and its application. IEEE Trans Ind Electron 64(6):5025–5034

Shen QK, Shi P, Zhu JW, Wang S, Shi Y (2019) Neural networks based distributed adaptive control of nonlinear multi-agent systems. IEEE Trans Neural Netw Learn Syst 31(3):1010–1021

Starodubtsev YN, Kataev VA, Bessonova KO, Tsepelev VS (2019) Hysteresis losses in nanocrystalline alloys with magnetic-field-induced anisotropy. J Magn Magn Mater 479(1):19–26

Su C-Y, Stepanenko Y, Svoboda I, Leung TP (2000) Robust adaptive control of a class of nonlinear systems with unknown backlash-like hysteresis. IEEE Trans Autom Control 12(45):2427–2432

Tao G, Kokotovic PV (1995) Adaptive control of plant with unknown hysteresis. IEEE Trans Autom Control 2(40):200–212

Tao Y-D, Li H-X, Zhu L-M (2019) Rate-dependent hysteresis modeling and compensation of piezoelectric actuators using Gaussian process. Sen Actuators A: Phys 295(15):357–365

Wang LX (1993) Stable adaptive fuzzy control of nonlinear systems. IEEE Trans Fuzzy Syst 1(2):146–155

Wang LX (1994) Adaptive fuzzy systems and control-design and stability analysis. Prentice Hall, NJ

Zhang TP (2002) Aadptive fuzzy sliding mode control based on a modified Lyapunov function. Acta Autom Sinica 28(1):137–142

Zhang TP (2002) On indirect adaptive fuzzy controller for a class of nonlinear systems. Control Decis 17(2):199–202

Chapter 5
Distributed Fuzzy Fault-Tolerant Output Consensus Control Against Unknown Nonlinear Dead-Zone

5.1 Introduction

For the leader-following systems considered in this chapter and the other class of multi-agent systems (MASs) including leader-following systems (LFSs), in the last two decades, their consensus problems always are important problems in control field, and extensive attention is drawn from exporters and scholars Chen et al. (2010), Murray (2007), Ren and Cao (2011), Qu (2009), Olfati-Saber et al. (2007), Wang et al. (2015), Szupiluk et al. (2014), Sedziwy (2014). For this study, leader-following consensus has sparked an increasing interest Shen et al. (2014), Shen and Shi (2015), Shi and Shen (2015), Li et al. (2013), Zhu and Cheng (2010), Hong et al. (2008), Yu and Xia (2012), Ren et al. (2007), Zhang and Lewis (2012), Mathiyalagan et al. (2015), Mathiyalagan et al. (2016), Mathiyalagan et al. (2015), Hu et al. (2013). In general, the consensus includes output and state consensus. At the first stage of this research, however, in the results reported in literature, only linear MASs were focused. For nonlinear MASs, only simple nonlinear MASs, such as first- or second-order MAS, were investigated. In recent years, researchers began to investigate the so-called high-order nonlinear MASs Shen et al. (2014), Shen and Shi (2015), Shi and Shen (2015). However, the results such as Shen et al. (2014), Shen and Shi (2015), Shi and Shen (2015) need a specific condition that the actuator is healthy, which means that fault or input nonlinearities described in the previous chapters are not occurred in real engineering. What's more, for system performance, it will be decreased by the nonlinearities or faults. Therefore, for the nonlinear MASs including leader-following systems subjected to the actuator faults and nonlinearities, their consensuses become necessary and important, and the relevant compensation methods should be proposed to reduce or remove their effects, which firstly motivate me for this work.

As common input nonlinearities, dead-zone and sector nonlinearity have widely been investigated by the exporters from various fields in the past 10 years. To reduce or even remove the influences of the nonlinearities, a large number of scholars and

Q. Shen, *Distributed Fault-Tolerant Consensus Control of Leader-Following Systems*, Intelligent Control and Learning Systems 11,
https://doi.org/10.1007/978-981-99-7426-9_5

engineering technicians made considerable efforts Hsu et al. (2004), Yan et al. (2005), Shyu et al. (2005), Hua et al. (2008), Hua and Ding (2011), Zhang and Ge (2009), Tong and Li (2012), Tong and Li (2013), Jiang et al. (2012). Some constraints, however, exist in the results. In Hsu et al. (2004); Yan et al. (2005); Shyu et al. (2005), all the parameters in sector nonlinearity must be known in advance. In Hsu et al. (2004), for the lumped uncertainty, it is assumed that its parameters are known exactly by the designers. Unfortunately, because all the prior knowledge cannot be accessed, such assumption and condition do not always hold, which means that the condition and assumption are not satisfied in the practical application, which is the second motivation of this work.

For the above systems with nonlinearities Hsu et al. (2004), Yan et al. (2005), Shyu et al. (2005), Zhou et al. (2006), Hua et al. (2008), Hua and Ding (2011), Zhang and Ge (2009), Tong and Li (2012), Tong and Li (2013), Li et al. (2015), in fact, the researchers have made considerable efforts to handle them. However, in the existing results, the systems considered in Hua et al. (2008), Hua and Ding (2011), Zhang and Ge (2009), Tong and Li (2012) belong to single-agent systems (SASs). The control methods in Zhou et al. (2006), Hua et al. (2008), Hua and Ding (2011), Zhang and Ge (2009), Tong and Li (2012), Tong and Li (2013) cannot be directly extended to the leader-following systems investigated in this chapter and other MASs because the multi-agent systems such as leader-following systems are more complex than SASs. Furthermore, for stability analysis and controller designing, there are clear and significant differences between MASs and SASs. What's important, multi-agent systems have wider application fields than SASs.

In this chapter, for a class of leader-following systems, their output consensus controls subjected to nonlinear unknown dead-zone are studied, where the followers are simultaneously subject to dead-zone and sector nonlinearity. A distributed adaptive fuzzy control method has been designed to guarantee that the output synchronization between the leader and followers is obtained and the tracking errors converge asymptotically to an adjustable neighborhood of the origin.

5.2 Preliminaries and Problem Formulation

In this section, basic graph theory is first introduced, and the consensus control problem is formulated for a class of leader-following system. In addition, the mode of nonlinear dead-zone is given in this section. Then, fuzzy logic systems are described.

5.2.1 Basic Graph Theory and Preliminaries

As stated in Chap. 3, in the sense of mathematics, MASs including LFSs investigated in this book can be represented by a graph. Obviously, it involves basic graph theory in the research of MASs. So, let us first introduce it in this subsection. For basic graph theory, it was already introduced in the previous chapter. For the convenience of reading, its related knowledge will be re-introduced here again. If the readers have gained an understanding of the basic graph theory, the readers can skip this part and read the following content directly.

A class of multi-agent systems will be considered here, which consist of N followers and a leader. For the leader and followers, the information exchange among the leader and followers is often modeled as a graph Zhang and Lewis (2012). Hence, we first introduce the knowledge of basic graph theory in this subsection.

A directed graph can be represented as $\mathscr{G} = (\mathscr{V}, \mathscr{E})$, where agent set $\mathscr{V}$ and edge (arcs) set $\mathscr{E}$ are defined as $\mathscr{V} = (\mathscr{V}_1, \ldots, \mathscr{V}_N)$ and $\mathscr{E} \subseteq \mathscr{V} \times \mathscr{V}$, respectively. Note that, if there exists an edge from node i to node j, then $(\mathscr{V}_i, \mathscr{V}_j) \in E$. For $\mathscr{G}$, its adjacency matrix $\mathscr{A}$ can be described as $\mathscr{A} = [a_{ij}] \in R^{N\times N}$, where element a_{ij} is defined as follows: (1) $a_{ij} = 0$ if there exists no an edge from node i to node j, i.e., $(\mathscr{V}_i, \mathscr{V}_j) \notin \mathscr{E}$ and (2) $a_{ij} \neq 0$ if there exists no an edge from node i to node j, i.e., $(\mathscr{V}_i, \mathscr{V}_j) \in E$. Throughout this chapter, a_{ii} is assumed to be equal to zero, i.e.,$a_{ii} = 0$, and the topology is also assumed to be fixed, which means that $\mathscr{A}$ is fixed and is not variant.

In directed graph $\mathscr{G}$, for the ith agent, its in-degree d_i is defined as $d_i = \sum_{j=1}^{N} a_{ij}$. Correspondingly, for the directed graph $\mathscr{G}$, its in-degree matrix and Laplacian matrix, i.e., $\mathscr{D}$ and $\mathscr{L}$, are also defined, respectively, as $\mathscr{D} = diag\{d_1, \ldots, d_N\} \in R^{N\times N}$ and $\mathscr{L} = [l_{ij}] = \mathscr{D} - \mathscr{A} \in R^{N\times N}$. For the Laplacian matrix $\mathscr{L}$, $L\underline{1} = 0$ with $\underline{1} = [1, \ldots, 1]^T \in R^{N\times 1}$. In addition, for agent i, its neighbor set can be described as $N_i = \{j|(\mathscr{V}_j, \mathscr{V}_i) \in E\}$. Obviously, for the ith agent, if the $j-th$ agent is its neighbor, agent i can obtain the information from the $j-th$ agent. For directed graph, however, not necessarily vice versa. For undirected graph, neighbors are interrelated. From node i to node j, if there exists a sequence of successive arcs, which can be described in the form $\{(\mathscr{V}_i, \mathscr{V}_l), (\mathscr{V}_l, \mathscr{V}_k), \ldots, (\mathscr{V}_m, \mathscr{V}_j)\}$, then a direct path exists. In a digraph, if there exist a special node and a directed path from the special node to every other node, then graph is said to has a spanning tree. For any two different nodes $\mathscr{V}_i$ and $\mathscr{V}_j$ in the digraph, from $\mathscr{V}_i$ to $\mathscr{V}_j$, if there always exists a direct path, then the graph is said to be strongly connected.

Let $\bar{\mathscr{E}} \subseteq \bar{\mathscr{V}} \times \bar{\mathscr{V}}$ and $\bar{\mathscr{V}} = \{\mathscr{V}_0, \mathscr{V}_1, \mathscr{V}_2, \ldots, \mathscr{V}_N\}$. Then, based on the graph $\mathscr{G}$ and a leader node, a new graph $\bar{\mathscr{G}}$ is defined as $\bar{\mathscr{G}} = \{\bar{\mathscr{V}}, \bar{\mathscr{E}}\}$. Note that, from the leader node to node k, if there exists an edge, then its weight $b_k > 0$, otherwise $b_k = 0$, where $k = 1, 2, \ldots, N$.

Now, the following lemma and assumption are made.

Lemma 5.1 (Zhang and Lewis 2012) *Define* $q = (\mathscr{B} + \mathscr{L})^{-1}\underline{1} = [q_1, \ldots, q_N]$,

$$P = diag\{\frac{1}{q_i}\} \in R^{N\times N} \ \ with \ \ p_i = \frac{1}{q_i}, \quad Q = P(\mathscr{B} + \mathscr{L}) + (\mathscr{B} + \mathscr{L})^T P,$$

then matrices Q *and* P *are positive definite, i.e.,* $Q > 0$ *and* $P > 0$.

Assumption 5.1 There exists a spanning tree in graph $\bar{\mathscr{G}}$, where the leader is the root, and its topology is invariant.

5.2.2 Problem Formation

A system has $N + 1$ nodes. Among the nodes, N nodes are regarded as followers, and the remaining node is seen as leaders. The followers are labeled $1, 2, \ldots, N$, and the leader is labeled 0. For the followers, their dynamics can be represented as

$$\begin{cases} \dot{v}_{k,1}(t) = v_{k,2}(t) \\ \dot{v}_{k,2}(t) = f_k(\bar{v}_k) + u_k(t) + \delta_k(\bar{v}_k, t) \\ y_k(t) = v_{k,1}(t), \end{cases} \tag{5.1}$$

where control input $u_k \in R$ and y_k denotes output; Function $f_k(\cdot) \in R$ is continuous and unknown; and the lumped disturbance $\delta_k(\bar{v}_k, t) \in R$ is bounded and unknown with $\bar{v}_k = (v_{k,1}, v_{k,2})^T$; $k = 1, 2, \ldots, N$. The leader has the following form:

$$\begin{cases} \dot{v}_{0,1}(t) = v_{0,2}(t) \\ \dot{v}_{0,2}(t) = f_0(\bar{v}_0, t) \\ y_0(t) = v_{0,1}(t), \end{cases} \tag{5.2}$$

where $v_{0,1} \in R$, $v_{0,2} \in R$, and y_0 are the leader's states and output, respectively; vector $\bar{v}_0(t) = (v_{0,1}, v_{0,2})^T \in R^2$; $f_0(\cdot) \in R$ denotes a bounded and unknown function.

For the kth follower, neighborhood output synchronization error is defined as

$$e_k(t) = \sum\nolimits_{j \in \mathcal{N}_k} a_{kj}(y_j(t) - y_k(t)) + b_k(y_0(t) - y_k(t)),$$

where the neighbors are described by $\mathcal{N}_k$ for the kth follower, weights $a_{k,j}$ and b_k are defined in the previous subsection, $k = 1, \ldots, N$. In addition, for the k follower, tracking error is also defined as

$$\varpi_k(t) = y_k(t) - y_0(t), \quad k = 1, \ldots, N.$$

In the real engineering, however, dead-zone often occurs in the actuators, whose detailed description is given later. In this chapter, it is assumed that the actuators in the followers have dead-zone property. If the actual output of the actuator subjected to dead-zone is described as $\varphi_k(u_k(t))$ with u_k being the designed input, then (5.1) becomes the following form:

$$\begin{cases} \dot{v}_{k,1}(t) = v_{k,2}(t) \\ \dot{v}_{k,2}(t) = f_k(\bar{v}_k) + I_k(u_k(t)) + \delta_k(\bar{v}_k, t) \\ y_k(t) = v_{k,1}(t). \end{cases} \tag{5.3}$$

Therefore, *the control objective* in this chapter is, designing an adaptive distributed control scheme to guarantee that the output synchronization between the leader and followers can be achieved regardless of the influence of the dead-zone.

Notice that, a graph $\bar{\mathcal{G}}$ is considered in this chapter as $\bar{\mathcal{G}} = \{\bar{\mathcal{V}}, \bar{\mathcal{E}}\}$, where the node set $\bar{\mathcal{V}}$ contains all the leader and followers, i.e., $\bar{\mathcal{V}} = \{\mathcal{V}_0, \mathcal{V}_1, \ldots, \mathcal{V}_N\}$, and the edge (arcs) set $\bar{\mathcal{E}}$ is defined as $\bar{\mathcal{E}} \subseteq \bar{\mathcal{V}} \times \bar{\mathcal{V}}$. The detailed description can be found in the previous subsection.

Now, let us introduce the following two assumptions and two valuable lemmas.

Assumption 5.2 The function $f_0(\bar{v}_0)$ is bounded, namely, $|f_0(\bar{v}_0, t)| \leq M_{f_0}$ with M_{f_0} being a known positive real constant.

Assumption 5.3 For follower k, the lumped disturbance $\delta_k(t)$ is bounded, namely, $|\delta_k(t)| \leq M_{\delta,k}$ with $M_{\delta,k}$ being an unknown positive real constant.

Lemma 5.2 (Zhang and Lewis 2012) *For* $i = 1, 2, \ldots, n$, $||\varpi_i|| \leq \frac{||e_i||}{\underline{\varsigma}(B+L)}$, *where* $\underline{\varsigma}(B + L)$ *is a minimum singular value of matrix* $B + L$.

Lemma 5.3 (Polycarpou and Ioannou 1995) *For* $\forall v \in R$, $|v| \leq \tanh(v/\delta)v + 0.2785\delta$ *with* $\delta > 0 \in R$.

5.2.3 Description of Dead-Zone

In this section, the dead-zone model investigated in this chapter is first described. From Fig. 5.1, it can be presented in the following form:

$$I(u) = \begin{cases} I_-(t)(u + u_L), & u < -u_L \\ 0, & -u_L \leq u \leq u_R \\ I_+(t)(u - u_R), & u > u_R, \end{cases} \tag{5.4}$$

where $I(u)$ is the output of the dead-zone while $u \in R$ denotes the input of the dead-zone; constants $u_L > 0$ and $u_R > 0$ are unknown; $I_-(t)$ and $I_+(t)$ are unknown functions, and the following conditions hold:

$$\begin{cases} (u + u_L)I_-(t)(u + u_L) \geq \tau_-(u + u_L)^2, u < -u_L \\ (u - u_R)I_+(t)(u - u_R) \geq \tau_+(u - u_R)^2, u > u_R \end{cases} \tag{5.5}$$

with real constants $\tau_- > 0$ and $\tau_+ > 0$ being bounded and unknown.

After analyzing (5.5), one further has

$$\begin{cases} I_-(u + u_L) \leq \tau_-(u + u_L), u < -u_L \\ I_+(u - u_R) \geq \tau_+(u - u_R), u > u_R. \end{cases} \tag{5.6}$$

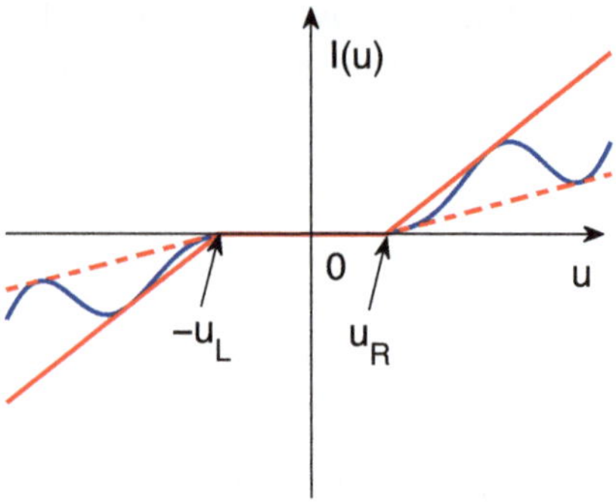

Fig. 5.1 Dead-zone function $I(u)$

Obviously,

$$\begin{cases} I_+ > 0, & u > \ \ u_R \\ I_- < 0, & u < -u_L. \end{cases}$$

Now, let the following symbols

$$\bar{u} = \max\{u_L, u_R\}, \quad \tau = \min\{\tau_+, \tau_-\}, \quad g' = \tau\bar{u}.$$

From the above description, it is easy to find that

$$\tau > 0, \quad \bar{u} > 0, \quad \text{and} \quad g' > 0.$$

Thus, from (5.5)–(5.6), it follows that

$$I(u) = \begin{cases} I_-(u + u_L) \leq \tau u + g', u < -u_L \\ 0, \quad -u_L \leq u \leq u_R \\ I_+(u - u_R) \geq \tau u - g', u > u_R. \end{cases} \tag{5.7}$$

In this chapter, each follower is subject to the above dead-zone model, and the dead-zone models in the followers are different. In other words, the parameters in dead-zones are different. Hence, subscript k is added to the symbols in the dead-zone. For example, for follower k, the symbols $I_k(u_k)$, $u_{k,+}$, $u_{k,-}$, $\tau_{k,+}$, $\tau_{k,-}$, $I_{k,+}$, and $I_{k,-}$ represent the dead-zone function and parameters. Notice that, for the function $I_k(u_k)$, it is obvious that (5.4)–(5.7) are satisfied.

Similarly, τ, $\bar{u}$, and b' are defined as follows:

$$\bar{u}_k = \max\{u_{k,L}, u_{k,R}\}, \quad \tau_k = \min\{\tau_{k,+}, \tau_{k,-}\}, \quad \text{and} \quad g_k' = \tau_k\bar{u}_k,$$

where $\bar{u}_k > 0$, $\tau_k > 0$, and $g_k' > 0$.

Now, an assumption is made for the above dead-zone.

Assumption 5.4 Unknown constants $\bar{u}_k$, τ_k are bounded, namely, $\tau_k \geq \tau_{\min}$ and $\bar{u}_k \leq u_{\max}$ with constants $u_{\min} > 0 \in R$ and $\tau_{\max} > 0 \in R$ being known.

5.2.4 *Fuzzy Logic Systems*

As stated in Chap. 3, as one of universal approximations, fuzzy logic systems (FLSs) have been utilized widely in control design of dynamic systems, especially uncertain nonlinear systems Shen and Shi (2015). The detailed description of FLSs is omitted here, which can be found in Chap. 3.

In this chapter, since $f_k(\bar{v}_k)$ in (1) is unknown, fuzzy logic system presented in Shen and Shi (2015) is adopted to approximate it as

$$f_k(\bar{v}_k) = \theta_k^{*T} S_k(v_k) + \varepsilon_k(v_k),$$

where $\varepsilon_k(v_k)$ denotes the optimal approximation error. Note that, in order to distinguish the functions in (5.1), the subscript k is added to the above symbols described FLSs.

From the universal approximation results of FLSs, an assumption is given.

Assumption 5.5 Ideal weight θ_k^* and optimal approximation error $\varepsilon_k(v_k)$ are bounded in a compact set, namely, $|\varepsilon_k(v_k)| \leq M_{\varepsilon,k}$ with unknown constant $M_{\varepsilon,k} > 0 \in R$.

5.3 Main Results

Now, for follower k, let us define control and adaptive laws as follows:

$$u_k = \begin{cases} -u_k' - u_{\max}, & z_{k,2} > 0 \\ 0, & z_{k,2} = 0 \\ u_k' + u_{\max}, & z_{k,2} < 0 \end{cases} \tag{5.8}$$

with

$$u_k' = \frac{|\hat{\theta}_k^T S_k| + |z_{k,2}| + \hat{g}_k + \hat{M}_{\varepsilon\delta,k} + |\dot{\alpha}_{k,1}|}{\tau_{\min}} \tag{5.9}$$

$$\dot{\hat{\theta}}_k = \begin{cases} \eta_1 z_{k,2} S_k, \ if \ ||\hat{\theta}_k|| < M_\theta \ or \\ \qquad ||\hat{\theta}_k|| = M_\theta \ and \ \hat{\theta}_k^T S_k \leq 0; \\ \eta_1 \hat{\theta}_k S_k - \eta_1 s \dfrac{\hat{\theta}_k \hat{\theta}_k^T}{||\hat{\theta}||^2} S_k, \\ \qquad if \ ||\hat{\theta}_k|| = M_\theta \ and \ \hat{\theta}_k^T S_k > 0 \end{cases} \tag{5.10}$$

$$\dot{\hat{M}}_{\varepsilon\delta,k} = \eta_2|z_{k,2}|, \ \hat{M}_{\varepsilon\delta,k}(0) > 0 \tag{5.11}$$

$$\dot{\hat{g}}_k = \eta_3|z_{k,2}|, \ \hat{b}_k(0) > 0 \tag{5.12}$$

with $\hat{\theta}_k$, $\hat{g}_k$, and $\hat{M}_{\varepsilon\delta,k}$ denote, respectively, the estimates of θ_k, g_k, and $M_{\varepsilon\delta,k}$; adaptive rates$\eta_1 > 0$, $\eta_2 > 0$, and $\eta_3 > 0$; and design parameter $M_\theta > 0$a.

Now, a theorem is given to show the detailed stability analysis.

Theorem 5.1 *Under Assumptions 5.1–5.5, consider leader-following system (5.1)–(5.3) with dead-zone (5.4). If control and adaptive laws in (5.8) and (5.9)–(5.12) are applied, then, for all followers, their outputs synchronize asymptotically to the leader's output, and each tracking error converges asymptotically to an adjustable set Ω_ϖ represented as*

$$\Omega_\varpi = \left\{ \varpi_k \mid \varpi_k \le \frac{\sqrt{2N \times 0.2785\delta_k}}{\underline{\varsigma}(L+B)}, \ k = 1, \ldots, N \right\}.$$

Proof In the following analysis, $k = 1, 2, \ldots, N$. Without causing ambiguity, the value of k is no longer given.

Since

$$e_k = \sum\nolimits_{j \in \mathcal{N}_k} a_{kj}(y_j - y_k) + b_k(y_0 - y_k),$$

then, from (5.3), one has

$$\begin{aligned} \dot{e}_k &= \sum\nolimits_{j \in \mathcal{N}_k} a_{kj}(v_{j,2} - v_{k,2}) + b_k(v_{0,2} - v_{k,2}) \\ &= d_k v_{j,2} - (b_k + d_k)v_{k,2} + b_k f_0 \end{aligned}$$

here $d_k = \sum_{j \in \mathcal{N}_k} a_{kj}$.

Notice that, $v_{k,2}$ in the above equation can be taken as a virtual control, which ensures $\lim_{t \to +\infty} e_k(t) = 0$.

Define the notations:

$$z_{k,1} = e_k, \ z_{k,2} = v_{k,2} - \alpha_{k,1}.$$

Define

$$V_1 = \frac{1}{2} Z_1^T Z_1 \ = \sum \frac{1}{2} z_{k,1}^2,$$

where

$$Z_1 = (z_{1,1}, \ldots, z_{N,1})^T.$$

Considering (5.8) and (5.9), the time derivative of V_1 can be described as follows:

$$
\begin{aligned}
\dot{V}_1 &= \sum_{k=1}^{N} \{z_{k,1}[d_k v_{j,2} - (b_k + d_k)v_{k,2} + b_k f_0] \\
&= \sum_{k=1}^{N} \{z_{k,1}[d_k v_{j,2} - (b_k + d_k)\alpha_{k,1}] + b_k f_0 - \\
&(b_k + d_k) z_{k,1} z_{k,2})\}.
\end{aligned} \tag{5.13}
$$

From Lemma 5.3 and Assumption 5.2, it follows that

$$
\begin{aligned}
z_{k,1} b_k f_0 &\le |b_k M_{f_0}| z_{k,1} \\
&\le z_{k,1} b_k M_{f_0} \tanh(\frac{b_k z_{k,1} M_{f_0}}{\delta_k}) + 0.2785\rho_k,
\end{aligned}
$$

where $\rho_k > 0$ is a design parameter.

By using Young's inequality, it can be further obtained

$$
\frac{1}{2}(z_{k,1}^2 + z_{k,2}^2) \ge z_{k,1} z_{k,2}.
$$

Thus, substituting the above equation into (5.13), one has

$$
\begin{aligned}
\dot{V}_1 &\le \sum_{k=1}^{N} \{z_{k,1}[d_k v_{j,2} - (b_k + d_k)\alpha_{k,1}] - [\frac{z_{k,1}^2}{2} + \\
&\frac{z_{k,2}^2}{2} + z_{k,1} b_k M_{f_0} \tanh(\frac{z_{k,1} b_k M_{f_0}}{\delta_k}) + 0.2785\rho_k]\} \\
&= \sum_{k=1}^{N} \{z_{k,1}[d_k v_{j,2} + b_k M_{f_0} \tanh(\frac{z_{k,1} b_k M_{f_0}}{\delta_k}) + \\
&\frac{z_{k,1}}{2}] - z_{k,1}(d_k + b_k)\alpha_{k,1} + \frac{1}{2} z_{k,2}^2 + 0.2785\rho_k\}.
\end{aligned} \tag{5.14}
$$

Let us define $\alpha_{k,1}$

$$
\alpha_{k,1} = \frac{M_{f_0} b_k \tanh(\frac{z_{k,1} M_{f_0} b_k}{\rho_k}) + z_{k,1} + d_k v_{j,2}}{b_k + d_k}. \tag{5.15}
$$

From (5.14) and (5.15), one has

$$
\dot{V}_1 \le \sum_{k=1}^{N} (-\frac{1}{2} z_{k,1}^2 + 0.2785\rho_k + \frac{1}{2} z_{k,2}^2). \tag{5.16}
$$

Remark 5.1 Notice that, once $z_{k,2}$ is bounded, namely, $|z_{k,2}| \le \kappa_{k,1}$ with design parameter $\kappa_{k,1} > 0$ being real constant, from (5.16), we can obtain

$$
\dot{V}_1 \le \sum_{k=1}^{N} (-\frac{1}{2} z_{k,1}^2 + \frac{1}{2} \kappa_{k,1}^2 + 0.2785\rho_k).
$$

Let $\mu_{k,1} = \frac{1}{2}\kappa_{k,1}^2 + 0.2785\rho_k$, then we have

$$\dot{V}_1 \le \sum_{k=1}^{N} (-\frac{1}{2} z_{k,1}^2 + \mu_{k,1})$$

and

$$\frac{1}{2} z_{k,1}^2 \le \mu_{k,1} \Rightarrow |z_{k,1}| \le \sqrt{2\mu_{k,1}}.$$

Therefore, $z_{k,1}$ is bounded under the condition that $z_{k,2}$ is bounded.

Now, the second equation of (5.3) is investigated, and suitable control laws are designed, respectively, for the followers such that $z_{k,2}, k = 1, 2, \ldots, n$, are bounded. From $z_{k,2} = v_{k,2} - \alpha_{k,1}$ and (5.3), it follows

$$\dot{z}_{k,2} = \delta_k + I_k(u_k) + f_k - \dot{\alpha}_{k,1}.$$

Define

$$V_2 = \sum \frac{1}{2} z_{k,2}^2.$$

Considering (5.3), the time derivative of V_2 is described as

$$\begin{aligned} \dot{V}_2 &= \sum_{k=1}^{N} z_{k,2} \dot{z}_{k,2} \\ &= \sum_{k=1}^{N} z_{k,2} (\delta_k + I_k(u_k) + f_k - \dot{\alpha}_{k,1}). \end{aligned}$$

Since $f_k(\bar{v}_k)$ is unknown but continuous, it can be approximated by an FLS as

$$f_k(v_k) = \theta_k^{*T} S_k(v_k) + \varepsilon_k(v_k),$$

where θ_k^* and $\varepsilon_k(v_k)$ denote, respectively, the ideal weight vector and optimal approximation error. Replacing f_k by $\theta_k^{*T} S_k(v_k) + \varepsilon_k(v_k)$, the equation of $\dot{V}_2$ can be described as

$$\dot{V}_2 = \sum_{k=1}^{N} z_{k,2} (\theta_k^{*T} S_k + \varepsilon_k + I_k(u_k) + \delta_k - \dot{\alpha}_{k,1}).$$

Note that, $\varepsilon_k(v_k)$ and $S(v_k)$ have been abbreviated respectively as S_k and ε_k.

From Assumptions 5.3 and 5.5, we have

$$\begin{aligned} z_{k,2}\delta_k &\le M_{\delta,k} z_{k,2} sgn(z_{k,2}) \\ z_{k,2}\varepsilon_k &\le M_{\varepsilon,k} z_{k,2} sgn(z_{k,2}) \end{aligned}$$

and

$$\begin{aligned} \dot{V}_2 \le \sum_{k=1}^{N} & z_{k,2} (\theta_k^{*T} S_k + M_{\varepsilon,k} sgn(z_{k,2}) + \\ & M_{\delta,k} sgn(z_{k,2}) - \dot{\alpha}_{k,1} + I_k(u_k)). \end{aligned} \tag{5.17}$$

Recall u_k in (5.8) and (5.9) , namely,

$$u_k = \begin{cases} -u_k' - u_{\max}, & z_{k,2} > 0 \\ 0, & z_{k,2} = 0 \\ u_k' + u_{\max}, & z_{k,2} < 0 \end{cases}$$

and

$$u_k' = \frac{|\hat{\theta}_k^T S_k| + |z_{k,2}| + \hat{g}_k + \hat{M}_{\varepsilon\delta,k} + |\dot{\alpha}_{k,1}|}{\tau_{\min}}.$$

According to the definition of u_k, the analysis below is divided into two cases.

Case 1: $z_{k,2} < 0$

Because of $z_{k,2} < 0$, one has

$$u_k = u_{k,\max} + u_k'.$$

And because $u_{k,\max} = \max\{u_{k,R}, u_{k,L}\}$, one has

$$u_k > u_{k,R} > 0.$$

And it follows from (5.7) that

$$I_k(u_k) > -g_k' + u_k \tau_k > 0.$$

Thus, one has

$$\begin{aligned} I_k(u_k) z_{k,2} &\leq \tau_k u_k z_{k,2} - g_k' z_{k,2} \\ &= \tau_k u_k' z_{k,2} + \tau_k u_{k,\max} z_{k,2} - g_k' z_{k,2} \\ &\leq \tau_k u_k' |z_{k,2}| + g_k |z_{k,2}|, \end{aligned} \tag{5.18}$$

where $g_k = -g_k' + u_{k,\max}\tau_k$.

Case 2: $z_{k,2} > 0$

Because of $z_{k,2} > 0$, one has

$$u_k = -u_{k,\max} - u_k'.$$

And because $u_{k,\max} = \max\{u_{k,R}, u_{k,L}\}$, one has

$$u_k < -u_{k,L} < 0.$$

From (5.7), we have

$$I_k(u_k) \leq g_k' + u_k \tau_k < 0.$$

Thus, one has

$$
\begin{aligned}
z_{k,2} I_k(u_k) &\le \tau_k u_k z_{k,2} + g'_k z_{k,2} \\
&= \tau_k u'_k z_{k,2} + \tau_k u_{k,\max} z_{k,2} + g'_k z_{k,2} \\
&\le \tau_k u'_k z_{k,2} + g_k z_{k,2} \\
&\le \tau_k u'_k |z_{k,2}| + g_k |z_{k,2}|.
\end{aligned} \tag{5.19}
$$

From the analysis in **Cases 1** and **2**, it follows

$$
z_{k,2} I_k(u_k) \le \tau_k u'_k |z_{k,2}| + g_k |z_{k,2}|. \tag{5.20}
$$

Substituting (5.20) into (5.17), one has

$$
\begin{aligned}
\dot{V}_2 \le & \sum_{k=1}^{N} z_{k,2} [\theta_k^{*T} S_k + M_{\varepsilon,k} sgn(z_{k,2}) + \\
& M_{\delta,k} sgn(z_{k,2}) - \dot{\alpha}_{k,1} + u'_k \tau_k sgn(z_{k,2}) + g_k sgn(z_{k,2})] \\
\le & \sum_{k=1}^{N} z_{k,2} \begin{bmatrix} \theta_k^{*T} S_k + M_{\varepsilon\delta,k} sgn(z_{k,2}) - \dot{\alpha}_{k,1} + \\ u'_k \tau_k sgn(z_{k,2}) + g_k sgn(z_{k,2}) \end{bmatrix},
\end{aligned} \tag{5.21}
$$

where $M_{\varepsilon\delta,k} = M_{\varepsilon,k} + M_{\delta,k}$.

Define

$$
\tilde{\theta}_k = \theta_k^* - \hat{\theta}_k, \quad \tilde{M}_{\varepsilon\delta,k} = M_{\varepsilon\delta,k} - \hat{M}_{\varepsilon\delta,k}, \quad \tilde{g}_k = g_k - \hat{g}_k.
$$

In the following analysis, the key task is to derive the adaptive laws of $\hat{g}_k$, $\hat{M}_{\varepsilon\delta,k}$, and $\hat{\theta}_k$. Firstly, $\hat{\theta}_k^T S_k |z_{k,2}|$, $\hat{M}_{\varepsilon\delta,k} |z_{k,2}|$, $|\dot{\alpha}_{k,1}||z_{k,2}|$ and $\hat{g}_k |z_{k,2}|$ are added to and subtracted from the right side in (5.21). Then one has

$$
\begin{aligned}
\dot{V}_2 \le & \sum_{k=1}^{N} z_{k,2} (\tilde{\theta}_k^{*T} S_k - \dot{\alpha}_{k,1} + \tilde{M}_{\varepsilon\delta,k} sgn(z_{k,2}) + \\
& u'_k \tau_k sgn(z_{k,2}) + sgn(z_{k,2}) \tilde{g}_k - \\
& \hat{\theta}_k^T S_k + \hat{M}_{\varepsilon\delta,k} sgn(z_{k,2}) + \hat{g}_k sgn(z_{k,2})).
\end{aligned} \tag{5.22}
$$

Recall (5.10)–(5.12), namely,

$$
\dot{\hat{\theta}}_k = \begin{cases} \eta_1 z_{k,2} S_k, \; if \; ||\hat{\theta}_k|| < M_\theta \; or \\ \qquad ||\hat{\theta}_k|| = M_\theta \; and \; \hat{\theta}_k^T S_k \le 0; \\ \eta_1 \hat{\theta}_k S_k - \eta_1 s \dfrac{\hat{\theta}_k \hat{\theta}_k^T}{||\hat{\theta}||^2} S_k, \\ \qquad if \; ||\hat{\theta}_k|| = M_\theta \; and \; \hat{\theta}_k^T S_k > 0, \end{cases}
$$

$$
\dot{\hat{M}}_{\varepsilon\delta,k} = \eta_2 |z_{k,2}|, \; \hat{M}_{\varepsilon\delta,k}(0) > 0,
$$

$$
\dot{\hat{g}}_k = \eta_3 |z_{k,2}|, \; \hat{b}_k(0) > 0.
$$

Define

$$V = V_1 + V_2 + \sum_{k=1}^{N} \left(\frac{\tilde{\theta}_k^T \tilde{\theta}_k}{2\eta_1} + \frac{\tilde{M}_{\varepsilon\delta,k}^2}{2\eta_2} + \frac{\tilde{g}_k^2}{2\eta_3}\right). \tag{5.23}$$

Differentiating V with respect to time t, it yields

$$\dot{V} = \dot{V}_1 + \dot{V}_2 - \sum_{k=1}^{N} \left(\frac{\tilde{\theta}_k^T \dot{\hat{\theta}}_k}{\eta_1} + \frac{\tilde{M}_{\varepsilon\delta,k} \dot{\hat{M}}_{\varepsilon\delta,k}}{\eta_2} + \frac{\hat{g}_k \dot{\hat{g}}_k}{\eta_3}\right). \tag{5.24}$$

Substituting (5.10)–(5.12), (5.16) and (5.22) into (5.24), it follows

$$\dot{V} \leq \sum_{k=1}^{N} \left(-\frac{z_{k,1}^2}{2} - \frac{z_{k,2}^2}{2} + 0.2785\rho_k\right) \tag{5.25}$$

which implies that, if

$$\sum_{k=1}^{N} (z_{k,1}^2 + z_{k,2}^2) > N \times 0.2785\rho_k$$

then

$$\dot{V} < 0.$$

Hence, we have

$$\sum_{k=1}^{N} |z_{k,1}| \leq \sqrt{2N \times 0.2785\rho_k}$$

is an invariant set. That is, if

$$\sum_{k=1}^{N} |z_{k,1}(0)| \leq \sqrt{2N \times 0.2785\rho_k}$$

then

$$\sum_{k=1}^{N} |z_{k,1}(t)| \leq \sqrt{2N \times 0.2785\rho_k}, \ \forall t \geq 0.$$

Since

$$Z_1 - (z_{1,1}, \ldots, z_{N,1})^T$$

then, from (5.3) and (5.9), we have

$$Z_1 = (e_{1,1}, \ldots, e_{N,1})^T = (B + L)y - \underline{y}_0,$$

where $y = (y_1, \ldots, y_N)^T$ and $\underline{y}_o = (y_0, \ldots, y_0)^T \in R^{N \times 1}$. Then, we have

$$||Z_1|| \leq \sqrt{2N \times 0.2785\rho_k}.$$

Further, from Lemma 5.2, one has

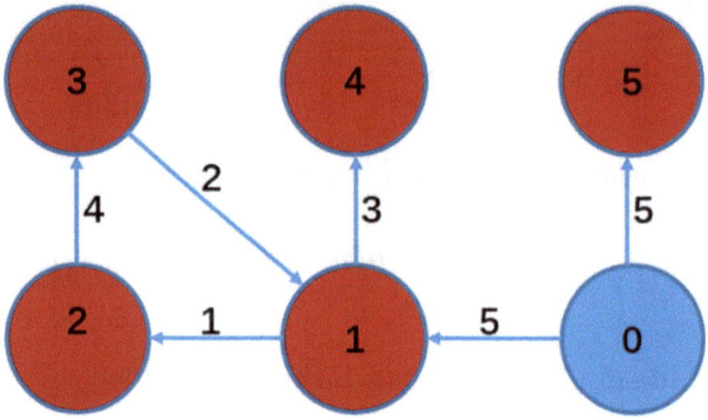

Fig. 5.2 Communication topology $\overline{G}$

$$|\varpi_k| \leq \frac{\sqrt{2N \times 0.2785\rho_k}}{\underline{\varsigma}(B+L)}.$$

This implies, for the followers in graph, their outputs synchronize asymptotically to the leader's output, and each tracking error converges asymptotically to an adjustable set, i.e.,

$$\begin{aligned}\varpi_k &\in \Omega_{\varpi} \\ &= \left\{\varpi_k | \varpi_k \leq \frac{\sqrt{2N \times 0.2785\rho_k}}{\underline{\varsigma}(B+L)}, k = 1, \ldots, N\right\}.\end{aligned} \tag{5.26}$$

The proof is completed.

5.4 Simulation

In order to show the effectiveness of the control strategy in this chapter, by using a numerical example, some simulation results are given. In this simulation, let us investigate a leader-following system. The system contains five followers and a leader, which is described by a digraph in Fig. 5.2. In the graph, a circle represents a node, the numbers inside the circles represent the labels of the nodes, a number on an edge that connects two nodes represents their weight, i.e., b_k and a_{ij}.

For leader labeled 0, it has the following form:

$$\begin{cases} \dot{v}_{0,1} = v_{0,2} \\ \dot{v}_{0,2} = -v_{0,2} + 3\sin(2t) + 6\cos(2t) - \frac{1}{3}(v_{0,1} + v_{0,2} - 1)^2 \cdot (v_{0,1} + 4v_{0,2} - 1) \\ y_0 = v_{0,1}. \end{cases}$$

Similar to (5.3), followers can be presented as

$$\begin{cases} \dot{v}_{k,1} = v_{k,2} \\ \dot{v}_{k,2} = f_k(\bar{v}_k) + I_k(u_k) + \delta_k(t) \\ y_k = v_{k,1}, \end{cases}$$

where $k = 1, 2, 3, 4$,

$$f_1(\cdot) = v_{1,2}\sin(v_{1,1}) + \cos(v_{1,1})^2$$

$$f_2(\cdot) = v_{2,1}v_{2,2}$$

$$f_3(\cdot) = v_{3,1}v_{3,2} + 0.2v_{3,2}^2$$

and

$$f_4(\cdot) = v_{4,1}.$$

For the dead-zone, in this work, the parameters are assumed that, for $k = 1, 2, 3, 4$,

$$I_{k,-} = 1 - 0.5\cos(t)$$

$$I_{k,+} = 1 + 0.5\sin(t)$$

$$u_{k,R} = 0.5, \quad u_{k,L} = 0.4.$$

It is easy to see that Assumption 4 is satisfied. Further, it can be obtained that $\tau_{\min} = 0.5$ and $u_{\max} = 0.5$. In this simulation, the disturbance $\delta_k (k = 1, \ldots, 4)$ takes an arbitrary value on the interval $[-1, 1]$. Obviously, it is bounded and $|\delta_k| \leq 1$. The sampling time is 0.08 s.

The initial states are chosen as

$$v_0 = [0, 0]^T, \; v_1 = [4, -3]^T$$

$$v_2 = [3, -2]^T, \; v_3 = [-3, -1]^T, \; v_4 = [-4, 2]^T.$$

Design parameters are set as

$$\eta_k = 2, \; \varepsilon_k = 0.01, \quad k = 1, 2, 3, 4.$$

In addition, it is easy to obtain from Fig. 5.2 that $B = diag\{5, 0, 0, 0\}$.

Some simulation results are given in Figs. 5.3, 5.4, and 5.5. From Fig. 5.3, we find easily, the output's synchronization between the leader and followers has been achieved. It is also found from Fig. 5.4 that each output tracking error converges asymptotically to a small neighborhood of the origin, which means that good consensus performance has been achieved under the control scheme defined by (5.8)–

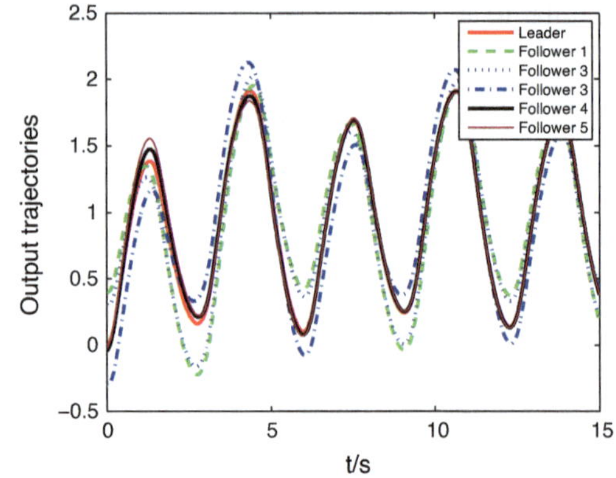

Fig. 5.3 The outputs of the leader and followers

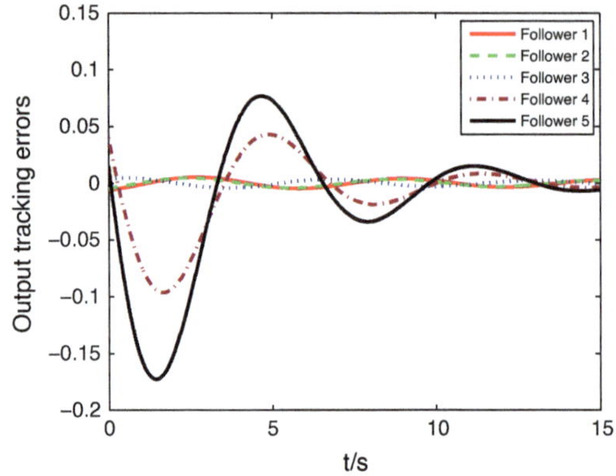

Fig. 5.4 The output tracking errors of the followers

(5.12). In addition, the boundedness of control signals is demonstrated in Fig. 5.5. The simulation results show clearly the effectiveness of the control scheme presented in this chapter.

5.5 Conclusion

In this chapter, the output consensus control problem is studied for a class of uncertain nonlinear leader-following systems on directed graph. The topology is assumed to be fixed, and all followers are subject to the unknown dead-zone input. By combining adaptive control with fuzzy logic systems, distributed fuzzy adaptive control scheme is constructed to ensure output synchronization between the leader and followers with the output tracking errors converging to an adjustable neighborhood of the origin.

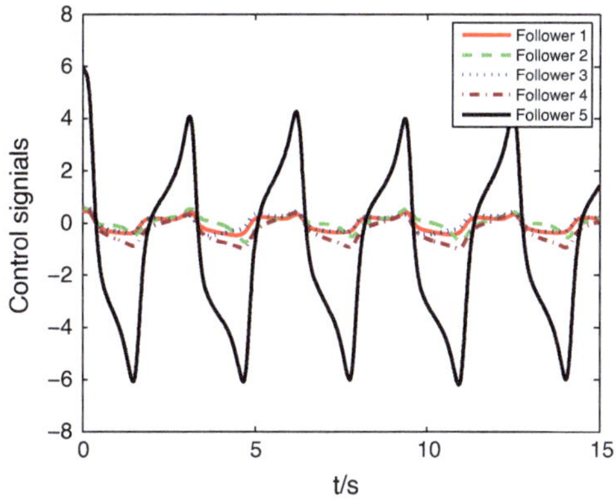

Fig. 5.5 The control signals of the followers

References

Chen J, Cao X, Cheng P, Xiao Y, Sun Y (2010) Distributed collaborative control for industrial automation with wireless sensor and actuator networks. IEEE Trans Industr Electron 57(12):4219–4230

Murray R (2007) Recent research in cooperative control of multivehicle systems. ASME J Dyn Syst Meas Control 129(5):571–583

Ren W, Cao C (2011) Distributed coordination of multi-agent networks: emergent problems, models, and issues. Springer, London

Qu Z (2009) Cooperative control of dynamical system: application to autonomous vehicles. Springer, London

Olfati-Saber R, Alex Fax J, Murray R (2007) Consensus and cooperation in networked multi-agent systems. Proc IEEE 95(1):215–233

Shen QK, Jiang B, Shi P (2014) Cooperative adaptive fuzzy tracking control for networked unknown nonlinear multi-agent systems with time-varying actuator faults. IEEE Trans Fuzzy Syst 22(3):494–504

Shen QK, Shi P (2015) Distributed command filtered backstepping consensus tracking control of nonlinear multiple-agent systems in strict-feedback form. Automatica 53:120–124

Shi P, Shen QK (2015) Cooperative control of multi-agent systems with unknown state-dependent controlling effects. IEEE Trans Autom Sci Eng 12(3):827–834

Li S, Du H, Shi P (2013) Distributed attitude control for multiple spacecraft with communication delays. IEEE Trans Aerosp Electron Syst 50(3):1765–1773

Zhu W, Cheng D (2010) Leader-following consensus of second-order agents with multiple time-varying delays. Automatica 46(12):1994–1999

Hong Y, Chen G, Bushnell L (2008) Distributed observers design for leader-following control of multi-agent networks. Automatica 44(3):846–850

Yu H, Xia X (2012) Adaptive consensus of multi agents in networks with jointly connected topologies. Automatica 48(8):1783–1790

Ren W, Moore KL, Chen Y (2007) High-order and model reference consensus algorithms in cooperative control of multivehicle systems. ASME J Dyn Syst Meas Control 129(5):678–688

Zhang H, Lewis FL (2012) Adaptive cooperative tracking control of higher-order nonlinear systems with unknown dynamics. Automatica 48(7):1432–1439

Hsu KC, Wang WY, Lin PZ (2004) Sliding mode control for uncertain nonlinear systems with multiple inputs containing sector nonlinearities and deadzones. IEEE Trans Syst Man Cybernet-Part B: Cybernet 34(1):374–380

Yan JJ, Shyu KK, Lin JS (2005) Adaptive variable structure control for uncertain chaotic systems containing dead-zone nonlinearity. Chaos, Solitions & Fractals 25(2):347–355

Shyu KK, Liu WJ, Hsu K (2005) Design of large- scale time-delayed with dead-zone input via variable structure control. Automatica 41(7):1239–1246

Ryan EP (1991) A universal adaptive stabilizer for a class of nonlinear systems. Syst Control Lett 16:209–218

Polycarpou MM, Ioannou PA (1995) A robust adaptive nonlinear control design. Automatica 31(3):423–427

Wang H, Liao X, Huang T, Li C (2015) Cooperative distributed optimization in multiagent networks with delays. IEEE Trans Syst Man Cybernet: Syst 45(2):363–369

Zhou J, Wen C, Zhang Y (2006) Adaptive output control of nonlinear systems with uncertain dead-zone nonlinearity. IEEE Trans Autom Control 51(3):504–511

Hua CC, Wang QG, Guan XP (2008) Adaptive tracking controller design of nonlinear systems with time delays and unknown dead-zone input. IEEE Trans Autom Control 53(7):1753–1759

Hua CC, Ding SX (2011) Model following controller design for large-scale systems with time-delay interconnections and multiple dead-zone inputs. IEEE Trans Autom Control 56(4):962–968

Zhang TP, Ge SS (2009) Adaptive neural network tracking control of MIMO nonlinear systems with unknown dead zones and control directions. IEEE Trans Neural Netw 20(3):483–497

Tong SC, Li YM (2012) Adaptive fuzzy output feedback tracking backstepping control of strict-feedback nonlinear systems with unknown dead zones. IEEE Trans Fuzzy Syst 20(1):168–180

Tong SC, Li YM (2013) Adaptive fuzzy output feedback control of MIMO nonlinear systems with unknown dead-zone inputs. IEEE Trans Fuzzy Syst 21(1):134–146

Li YM, Tong SC, Li TS (2015) Observer-based adaptive fuzzy tracking control of MIMO stochastic nonlinear systems with unknown control direction and unknown dead-zones. IEEE Trans Fuzzy Syst 23(4):1228–1241

Mathiyalagan K, Park JH, Sakthivel R (2015) Observer-based dissipative control for networked control systems: A switched system approach. Complexity 21(2):297–304

Mathiyalagan K, Park JH, Sakthivel R (2016) Robust passivity based resilient control for networked control systems with random gain fluctuations. Int J Robust Nonlinear Control 26(3):426–444

Mathiyalagan K, Park JH, Sakthivel R (2015) Non-fragile observer-based H∞ control for discrete-time systems using passivity theory. Circuits Syst Signal Process 34:2499516

Jiang HB, Zhang LP, Yu JJ, Zhou CG (2012) Consensus of multi-agent systems with dead-zone nonlinearity. Int J Control Autom Syst 10(4):824–829

Hu HX, YU L, Chen G, Xie GM (2013) Second-order consensus of multi-agent systems with unknown but bounded disturbance. Int J Control Autom Syst 11(2):258–267

Szupiluk R, Wojewnik P, Zabkowski T (2014) Smooth component analysis and MSE decomposition for ensemble methods in multi-agent environment. Int J Innov Comput Inf Control 10(4):1435–1445

Sedziwy A (2014) Effective graph representation supporting multi-agent distributed computing. Int J Innov Comput Inf Control 10(1):101–113

Chapter 6
Adaptive Fault-Tolerant Consensus Control Against Communication Faults

6.1 Introduction

For the last two decades, consensus control problem of multi-agent systems (MASs) has attracted extensive attention from the control community, and fruitful results have been obtained, see for example, Yu and Xia (2012), Du et al. (2012), Wang et al. (2007), Chen et al. (2010), Murray (2007), Zhang and Lewis (2012), Ren et al. (2007), Shen et al. (2014) and the references therein. Among MASs, leader-following systems (LFSs) belong to a special MASs, whose agents divide into two classes: leader(s) and followers. The number of leaders can be one or larger than one. For the leader-following systems (LFSs), there exist many research directions. Among the directions, consensus is an important issue Zhu and Cheng (2010), Ni and Cheng (2010), Yu and Wang (2010), Ren et al. (2007), Zhang et al. (2011), Shen and Shi (2015), Shi and Shen (2015). In literature, it is often called as synchronization problem. In general, the relevant control schemes are designed so that the followers can synchronize to a model reference signal or a leader Wang and Slotine (2006); Ren et al. (2007). In some literatures, it is also called pinning control or leader-following control Hong et al. (2008), Wang and Chen (2002). For such consensus control problem, in fact, the leaders act as a command generator to generate the desired reference signal and ignore the information from the followers in the whole networked systems. At the same time, for each follower in the network, only the information of itself and its neighbors can be utilized to construct distributed control scheme, which ensures that the synchronization between itself and the leader is achieved. In other words, each follower can obtain the local information from its neighbors via communication or communication network. In practical applications, among multiple agents, the information is transmitted via communication links. Communication is always subject to many constrains, for example, sensor noise or faults, channel fading, quantization errors, and so on Ni and Cheng (2010), Yu and Wang (2010), Ren et al. (2007), Zhang et al. (2011), Shen and Shi (2015), Shi and Shen (2015). Obviously, the constrains maybe result in the faults occurred in the information communication

Q. Shen, *Distributed Fault-Tolerant Consensus Control of Leader-Following Systems*, Intelligent Control and Learning Systems 11,
https://doi.org/10.1007/978-981-99-7426-9_6

links, which are named communication faults in this chapter. In the relevant results reported in literatures, however, the communication faults are seldom investigated, and only classical faults such as actuator and sensor faults are concerned Ni and Cheng (2010), Yu and Wang (2010), Ren et al. (2007), Zhang et al. (2011), Shen and Shi (2015), Shi and Shen (2015). Therefore, for the LFSs with the communication faults, how to construct a suitable distributed control method becomes naturally more important and challenging, which is the first motivation of the current research.

In recent years, for LFSs with the communication faults or other constrained conditions, leader-following consensus or synchronization control was investigated Consensus of discrete-time multi-agent systems with transmission nonlinearity (2013), Hu and Feng (2010), Huang and Manton (2009), Li and Zhang (2009), Zheng and Shen (2023). In Consensus of discrete-time multi-agent systems with transmission nonlinearity (2013), the consensus of discrete-time MASs with communication faults was discussed, where the assumptions imposed on the communication fault functions are acceptable although it is somewhat restrictive. In Hu and Feng (2010), under noise measurement, for a class of first-order leader-following systems represented in the discrete-time form, their distributed tracking control problem was investigated. However, the authors investigated only communication bias fault in Hu and Feng (2010). Considering noisy measurements, for leaderless MASs, Huang and Manton (2009) studied a consensus control problem and proved that the average consensus can be achieved under some given conditions. Note that, the MASs considered in Huang and Manton (2009) are discrete time. In Li and Zhang (2009), the authors extended the results obtained in Huang and Manton (2009) to continue-time systems. For linear or nonlinear multi-agent systems including LFSs considered in this chapter, under the condition that the communication faults have occurred, how to compensate for the communication faults and construct a suitable distributed control scheme, is very important. However, the research is not sufficient and needs further study, which is still an open topic.

In this chapter, we investigate the consensus of leader-following systems with communication faults and propose distributed adaptive control strategy such that the consensus between the leader and followers is achieved, meanwhile, the communication faults have been compensated. Compared with the existing works, our contributions are as follows:

(1) For the communication faults, two classes of faults are modeled firstly. Taking into account the fault mechanism and its effects, two fault models are proposed in this chapter. By using the fault models, many types of communication faults in actual engineering can be handled; and different time-varying.

(2) For the two types of communication faults, unlike Consensus of discrete-time multi-agent systems with transmission nonlinearity (2013), Hu and Feng (2010), Huang and Manton (2009), Li and Zhang (2009), Zheng and Shen (2023), two new adaptive distributed consensus control strategies are constructed respectively in this chapter. Meanwhile, the corresponding compensation algorithms are proposed for these two types of faults.

The rest of this chapter is organized as follows. The problem considered in this chapter is formulated and graph theory is recalled in Sect. 6.2. The main theoretical results in this chapter are given in Sect. 6.3. In Sect. 6.4, by numerical example, some simulation results are obtained, which show the effectiveness of the designed control strategy. Finally, Sect. 6.5 draws the conclusion.

Notations: Throughout this paper, R, R^n and $R^{n\times m}$ represent, respectively, the real numbers, the real n-vectors and the real $n \times m$ matrices; $|\cdot|$ is the absolute value of a real number; $||\cdot||$ is the Euclidean norm of a vector; $||\cdot||_F$ is the Frobenius norm of a matrix; $tr\{\cdot\}$ is the trace of a matrix; $\varsigma(\cdot)$ is the set of singular values of a matrix, with the maximum singular value $\bar{\varsigma}(\cdot)$ and the minimum singular value $\underline{\varsigma}(\cdot)$; $\lambda(\cdot)$ is the set of eigenvalues of a matrix, with the smallest eigenvalue $\lambda_{\min}(\cdot)$ and the largest eigenvalue $\lambda_{\max}(\cdot)$; matrix $P > 0$ ($P \geq 0$) means P is positive definite (positive semi-definite); I denotes the identity matrix with appropriate dimensions.

6.2 Problem Formulation and Preliminaries

In this section, some notations and basic graph theory are firstly introduced. Then, the consensus of leader-following systems is formulated. Finally, two communication fault models are presented.

6.2.1 *Notations and Basic Graph Theory*

Let $\mathcal{G} = (v, \mathcal{E})$ represent a graph, where sets $v = (v_1, \ldots, v_n)$ and $\mathcal{E} \subseteq v \times v$ represent respectively the sets of vertices and edges, $v_i, i = 1, 2, \ldots, n$, denotes the nodes in graph $\mathcal{G}$. If there exists an edge from node i to node j, then $(v_i, v_j) \in \mathcal{E}$; otherwise $(v_i, v_j) \notin \mathcal{E}$. For graph $\mathcal{G}$, matrix A is defined as $A = [a_{ij}] \in R^{n\times n}$, which is named as the adjacency matrix of $\mathcal{G}$. The elements of A, i.e., a_{ij} $(i, j = 1, 2, \ldots, n)$ denotes the weight of the edge from node i to node j, which is defined as follows:

Case 1: $i \neq j$

$$\begin{cases} a_{ij} > 0, & \text{if } (v_j, v_i) \in \mathcal{E} \\ a_{ij} = 0, & \text{if } (v_j, v_i) \in \mathcal{E} \end{cases}$$ In-degree matrix $D = diag\{d_1, \ldots, d_n\} \in R^{n\times n}$,

Case 2: $i = j$

$$a_{ii} = 0.$$

where $d_i = \sum_{j=1}^{n} a_{ij}$ is the weighted in-degree of node i. Graph Laplacian matrix $L = [l_{ij}] = D - A \in R^{n\times n}$, which has the following property: $L\underline{1} = 0$ with vector $\underline{1}$ being $\underline{1} = [1, \ldots, 1]^T \in R^n$.

In a graph, if node i can get the information from node j, then node j is a neighbor of node i. Note that, A node can access the information only from its own neighbors. For node i, its neighbor set is represented by $\mathcal{N}_i = \{j | (v_j, v_i) \in \mathcal{E}\}$.

Weight b_i is defined as follows:

$$\begin{cases} b_i > 0, & \text{if the leader is the neighbor of node } i \\ b_i = 0, & \text{if the leader is not the neighbor of node } i \end{cases}$$

Then, leader adjacency matrix B is defined as $B = diag\{b_1, \ldots, b_n\}$, where $b_i > 0$ if and only if the node i has access to the leader information; otherwise, $b_i = 0$.

Throughout this chapter, it is assumed that the topology is fixed, i.e., A and B are time-invariant.

Assumption 6.1 Undirected graph $\mathcal{G}$ is connected, and at least one follower has access to the information of the leader node 0.

Lemma 6.1 (Peng et al. 2013) *Under Assumption 6.1, and there exists at least one follower, which has access to the leader. Then* $B + L > 0$.

Lemma 6.2 (Zhang and Lewis 2012) *Define* $q = [q_1, \ldots, q_n] = (B + L)^{-1}\underline{1}$, $P = diag\{p_i\} = diag\{1/q_i\}$, $Q = P(B + L) + (B + L)^T P$, *then matrices* Q *and* P *are positive definite, i.e.,* $Q > 0$ *and* $P > 0$.

6.2.2 Problem Formulation

Consider the following $n(n \geq 2)$ followers. Follower i is described as:

$$\dot{z}_i = u_i \tag{6.1}$$

where $i = 1, \ldots, n$, $z_i \in R$ and $u_i \in R$ represent respectively state and input.

For the leader in the graph, it is labeled 0, which can be described as follows:

$$\dot{z}_0 = a(z_0, t) \tag{6.2}$$

where $z_0 \in R$ is the leader's state, $a(z_0, t)$ is assumed to unknown for all the followers in graph $\mathcal{G}$. In the following analysis, a symbol $\underline{z}_0$ will be used, which is defined as $\underline{z}_0 = z_0\underline{\mathbf{1}} \in R^n$ with $\underline{\mathbf{1}} = [1, \ldots, 1]^T \in R^n$.

In the existing works about leader-following systems Zhang and Lewis (2012), the leader (6.2) is often seen as a command generator, namely, the desired trajectory that the follower wants to track is generated by him. Notice that, the desired trajectory is dynamic if $a(z_0, t) \neq 0$, otherwise static.

In this chapter, the dynamics of the leader is assumed to be static, i.e., $\dot{z}_0 = 0$.

It is worthy to be emphasized that, each follower in the graph $\mathcal{G}$ only uses the lumped information in control designing. For example, follower i only use the lumped information ε_i, which is described as

$$\varepsilon_i = \sum\nolimits_{j \in N_i} a_{ij}(z_j - z_i) + b_i(z_0 - z_i) \tag{6.3}$$

where weights $b_i \geq 0$ is defined in the previous subsection. In the relevant literatures, ε_i is often called the *neighborhood synchronization error* of the ith follower.

Note that, $B = diag\{b_1, \ldots, b_n\} \in R^{n\times n}$. Define global synchronization error is defined as $\varepsilon = [\varepsilon_1, \varepsilon_2, \ldots, \varepsilon_n]^T \in R^n$, then one has

$$\varepsilon = -(L + B)(z - \underline{z}_0) \tag{6.4}$$

Since $L\underline{\mathbf{1}} = \underline{\mathbf{0}}$, from (6.1), one further has,

$$\dot{\varepsilon} = -(L + B)(u - \dot{\underline{z}}_0) \tag{6.5}$$

Now, for follower i $(i = 1, 2, \ldots, n)$, the following tracking error is defined

$$\delta_i = z_i - z_0.$$

Define $\delta = [\delta_1, \ldots, \delta_n]^T \in R^n$. Furthermore, one has

$$\delta = z - \underline{z}_0$$

where $\underline{z}_0 = [z_0, \ldots, z_0]^T \in R^n$.

For leader-following systems, consensus or synchronization control problem investigated in this chapter, the control objective is to design respectively control laws for followers in the graph $\mathcal{G}$ and asymptotical synchronization between the leader and followers can be achieved.

Notice that, as mentioned earlier, for each follower, only the information available is the information from itself and its neighbors. It is more important that the information is combined into one single signal ε_i in (6.3) Li et al. (2010).

To achieve the above control objective, in general, a distributed control law u_i is constructed for follower i in the following form:

$$u_i = f_i(\varepsilon_i) = M_{z_0} sgn(\varepsilon_i) b_i - h_i \varepsilon_i \tag{6.6}$$

where $h_i > 0$ is a design parameter, real constant $M_{z_0} > 0$ is the upper boundary of z_0, namely, $|z_0| \leq M_{z_0}$, $f_i(\varepsilon_i)$ is a suitable known function.

Define

$$V = \frac{1}{2}\varepsilon^T P \varepsilon,$$

where $P = P^T > 0$ is a matrix defined in Lemma 6.2. Now, differentiating it with respect to time t, it yields

$$\dot{V} = \varepsilon^T P \dot{\varepsilon}.$$

Substituting the control law defined by (6.6) into $\dot{V}$, from Lemmas 6.1 and 6.2, one has

$$\dot{V} \leq -h\varepsilon^T P(B+L)\varepsilon$$
$$= \frac{1}{2}h\varepsilon^T Q\varepsilon \qquad \leq 0,$$

where $h = diag\{h_1, \ldots, h_n\}$, and $Q > 0$.

Based on Lyapunov theory, it can be easily obtained that the neighborhood synchronization error ε_i is asymptotically bounded.

From Lemma 6.3, one can further obtain the tracking error δ_i is bonded too. It implies that the synchronization between the followers and leader has been achieved.

However, in communication networks, because of sensor faults, external disturbance, noise, and other factors, data transmission links might be faulty. It means that the data transmitted in these failed links is contaminated and the true value of the data is not available. As far as the work in this chapter is concerned, for ε_i, its true value can't be accessed by follower i. Hence, ε_i cannot be utilized in control design (6.6) to ensure that the synchronization is achieved. In this case, *communication faults* have occurred communication links. In the following subsection, the communication faults will be described in detail later.

Considering the communication faults, the main task in this chapter is twofold:

(I) Protocol is constructed for each follower to ensure that the synchronization between the leader and follower is achieved; and

(II) For the communication faults, their effect is investigated, and their influence on the system performance is revealed. Then, the faults are modeled, and the corresponding compensation algorithms are designed.

To design an appropriate consensus control scheme for the leader-following systems described by (6.1) and (6.2), the following lemma and assumption are made.

Lemma 6.3 (Zhang and Lewis 2012) $||\delta|| \leq \frac{||\varepsilon||}{\underline{\varsigma}(L+B)}$, *where* $\underline{\varsigma}(L+B)$ *is a minimum singular value of matrix* $L+B$, *where* $\varepsilon = [\varepsilon_1, \ldots, \varepsilon_n]^T$.

Notice that, for a class of leader-following systems with communication faults, this chapter considers their consensus control problem. As mentioned earlier, the main task is to construct a distributed control method for the leader-following system such that the synchronization control objective is achieved. Meanwhile, the influence of communication faults is relaxed by applying suitable compensation algorithms. Hence, in order to describe clearly the design process of the fault compensation algorithm and to reduce the complexity of derivation, the leader is assumed to be static.

Remark 6.1 It is well known that a follower in the graph accesses its neighbors' information. As done in Li et al. (2010), for follower i, the relative measurements of its neighbor nodes are synthesized into a single signal ε_i defined by (6.3). That is to say, only ε_i can be used to design the control input of follower i, and the states of its neighbors *cannot* be directly used for control design.

Remark 6.2 For consensus control of MASs including LFSs considered in this work, Assumption 1 is standard. As stated in Zhang and Lewis (2012), in the graph, if Assumption 1 does not be satisfied, then there is indeed a collection S_1, which contains l $(1 \leq l < n)$ followers. The other nodes form a subset S_2. The followers in the collection S_1 don't have a path to access the leader, which means that the l followers in S_1 don't obtain the information from the followers in S_2. Obviously, some nodes are isolated and cannot obtain the information from any other nodes. Hence, no suitable synchronization control method can be designed to guarantee that the synchronization between the followers and leader is achieved under this condition.

Remark 6.3 For the consensus problem of leader-following systems or MASs, there are abundant results in literature. However, most of these results focus on only the ideal case where the communication faults *do not* occur in signal communication. In this chapter, we will consider a more practical case: the consensus problem with communication faults.

6.2.3 Models of Transmission Faults

As stated previously, in communication network, data or signal transmission is affected by a variety of factors. The factors may result in the fact: for a piece of data, the resulting value is not equal to the true value of the data. For instance, for signal ε_i, the nodes in communication networks actually get through the communication network is $\varepsilon_{a,i}$. Obviously, if there exists no fault in communication links, then the actual value obtained by nodes, i.e., $\varepsilon_{a,i}$, should be equal to its true value, i.e., ε_i, namely, $\varepsilon_{a,i} = \varepsilon_i$. However, due to the presence of communication faults, they are not equal, namely, $\varepsilon_i \neq \varepsilon_{a,i}$.

Speaking generally, communication faults, according to the impact on data transmission, can be divided into the following two categories: (i) communication gain faults and (ii) communication bias faults.

Communication gain faults can be described as follows:

$$\varepsilon_{a,i} = \beta_i(\varepsilon_i, t) * \varepsilon_i, \quad \forall i \in \{1, \ldots, n\} \tag{6.7}$$

where $\beta_i(\varepsilon_i, t) > 0$ is bounded.

Communication bias fault has the following form:

$$\varepsilon_{a,i} = \varepsilon_i + w_i(\varepsilon_i, t), \quad \forall i \in \{1, \ldots, n\} \tag{6.8}$$

where $w_i(\varepsilon_i, t)$ is the external bounded signal.

In fault-free case, i.e., $\beta_i(\varepsilon_i, t) = 1$ and $w_i(\varepsilon_i, t) = 0$, so $e_{a,i} = \varepsilon_i$, the synchronization between the leader and followers can be guaranteed by the protocol given in (6.6). Due to the occurrence of the communication faults in real engineering, $\varepsilon_{a,l} \neq \varepsilon_i$, which implies that the control input in (6.6) cannot be designed by using $\varepsilon_{a,l}$, and further implies that the synchronization cannot be ensured by the control law in (6.6). Therefore, for the leader-following systems, considering the actual condition of the communication links, communication faults in account should be taken in their distributed control designing.

By using the available signal $\varepsilon_{a,i}$ and the fault modes described by (6.7) and (6.8), applying the same design approach in (6.6), the control law u_i in (6.6) can be re-designed as follows:

$$u_i = f_i(\varepsilon_{a,i}) \tag{6.9}$$

Because of $\varepsilon_i \neq \varepsilon_{a,i}$, $f_i(\varepsilon_{a,i}) \neq f_i(\varepsilon_i)$. Obviously, because of the communication faults, control law in (6.9) couldn't guarantee the synchronization although the control law in (6.6) could ensure the synchronization.

Therefore, in order to achieve the synchronization, the communication faults must be considered and compensated.

Now, an assumption is made for the communication faults.

Assumption 6.2 The faults and their first time derivatives, i.e., $\beta_i(\varepsilon_i, t)$, $\dot{\beta}_i(\varepsilon_i, t)$, $w_i(\varepsilon_i, t)$ and $\dot{w}_i(\varepsilon_i, t)$ are bounded and $0 < \beta_i(\varepsilon_i, t) < 1$, namely, there exist real constants $\bar{\beta}_i > 0$, $\underline{\beta}_i > 0$, $\bar{\bar{\beta}}_i > 0$, $\bar{w}_i > 0$ and $\bar{\bar{w}}_i > 0$ such that $\underline{\beta}_i(t) \leq \beta_i(\varepsilon_i, t) \leq \bar{\beta}_i$, $|\dot{\beta}_i(\varepsilon_i, t)| \leq \bar{\bar{\beta}}_i$, $|w_i(\varepsilon_i, t)| \leq \bar{w}_i$ and $|\dot{w}_i(\varepsilon_i, t)| \leq \bar{\bar{w}}_i$, $i = 1, \ldots, n$.

6.3 Main Results

In this section, two different distributed adaptive fault-tolerant control methods are designed, respectively, for the above two communication fault models in (6.7) and (6.8), which guarantee that the synchronization between the followers and leader. Note that, the two communication faults are taken into account and are compensated in the control design.

6.3.1 Fault-Tolerant Consensus Control Against Transmission Gain Faults

In this subsection, let us first consider the communication gain fault described by (6.7).

From the communication fault in (6.7), it follows

$$\varepsilon_{a,i} = \beta_i \varepsilon_i, \quad \varepsilon_i = \frac{1}{\beta_i} \varepsilon_{a,i} = \beta_{-1,i} \varepsilon_{a,i}$$

where

$$\beta_{-1,i} = \frac{1}{\beta_i}.$$

To achieve the synchronization, for each follower in the graph, the following control law is designed as

$$u_i = \frac{h_1 \hat{\beta}_{-1,i}}{(b_i + d_i)} \varepsilon_{a,i} \text{ or } u = h_1 \hat{\beta}_{-1} (B + D)^{-1} \varepsilon_a \tag{6.10}$$

where

$$u = [u_1, \ldots, u_n]^T, \quad \varepsilon_a = [\varepsilon_{a,1}, \ldots, e_{a,n}]^T,$$

$$\beta_{-1} = \begin{bmatrix} \beta_{-1,1} & & \\ & \ddots & \\ & & \beta_{-1,n} \end{bmatrix}, \quad \hat{\beta}_{-1} = \begin{bmatrix} \hat{\beta}_{-1,1} & & \\ & \ddots & \\ & & \hat{\beta}_{-1,n} \end{bmatrix}.$$

and $\hat{\beta}_{-1}$ is the estimate of β_{-1}, $\hat{\beta}_{-1,i}$ $(i = 1, \ldots, n)$ is the estimate of $\beta_{-1,i}$, $h_1 > 0$ is a design parameter satisfying the following condition:

$$\frac{h_1 \underline{\beta} \hat{\underline{\beta}}_{-1} \underline{\varsigma}(P) \underline{\varsigma}(L + B)}{\bar{\varsigma}(B + D)} - \bar{\beta}^2 (\bar{\hat{\beta}}_{-1} + \bar{\beta}_{-1}) \bar{\varsigma}(B + D) > \lambda_1 \tag{6.11}$$

where design parameter $\lambda_1 > 0$,

$$\underline{\hat{\beta}}_{-1} = \min\{\underline{\hat{\beta}}_{-1,1}, \ldots, \underline{\hat{\beta}}_{-1,n}\} \text{ and } \underline{\beta} = \min\{\underline{\beta}_1, \ldots, \underline{\beta}_n\}.$$

Note that, throughout this chapter, for a matrix, $\varsigma(\cdot)$ denotes one of the singular values with $\underline{\varsigma}(\cdot)$ and $\bar{\varsigma}(\cdot)$ being the minimum and maximum singular values.

For $\hat{\beta}_{-1,i}$, define its adaptive law

$$\dot{\hat{\beta}}_{-1,i} = proj\{\zeta_i\} = \begin{cases} 0, \text{ if } (\hat{\beta}_{-1.i} = \underline{\hat{\beta}}_{-1.i} \text{ or } \hat{\beta}_{-1.i} = \bar{\hat{\beta}}_{-1.i}) \\ \quad \text{and } \hat{\beta}_{-1.i}\zeta_i \geq 0 \\ \zeta_i, \text{ if } \underline{\hat{\beta}}_{-1.i} < \hat{\beta}_{-1.i} < \bar{\hat{\beta}}_{-1.i} \text{or } \hat{\beta}_{-1.i}\zeta_i < 0 \\ \quad \text{and } (\dot{\hat{\beta}}_{-1.i} = \underline{\dot{\hat{\beta}}}_{-1.i} \text{ or } \dot{\hat{\beta}}_{-1.i} = \bar{\dot{\hat{\beta}}}_{-1.i}) \end{cases}, \tag{6.12}$$

$$\underline{\hat{\beta}}_{-1.i} < \hat{\beta}_{-1.i}(0) < \bar{\hat{\beta}}_{-1.i}$$

where

$$\zeta_i = \eta_\beta \varepsilon_{a,i} (d_i + b_i) \varepsilon_{a,i} - \gamma_\beta \hat{\beta}_{-1,i},$$

adaptive rate $\eta_\beta > 0$ is a design parameter, the projection algorithm is applied to guarantee that $\hat{\beta}_i$ is bounded, namely, $\bar{\hat{\beta}}_{-1,i} \geq \hat{\beta}_{-1,i} \geq \underline{\hat{\beta}}_{-1,i}$ with $\underline{\hat{\beta}}_{-1,i} > 0$ and $\bar{\hat{\beta}}_{-1,i} > 0$ being design parameters.

Remark 6.4 Comparing (6.6) with (6.10), it can be easily seen, there is $\hat{\beta}_{-1,i}$ in (6.10). Note that, $\beta_{-1,i}$ will be compensated by $\hat{\beta}_{-1,i}$.

Remark 6.5 To remove or relax the influence of the communication gain fault β_i, it should be compensated in control design. In general, fault-tolerant control laws constructed for the followers have the following form:

$$u_i = \frac{h_1}{\hat{\beta}_i} \frac{1}{b_i + d_i} \varepsilon_{a,i} \text{ or } u = h_1 (B + D)^{-1} \hat{\beta}^{-1} \varepsilon_a$$

where $u = [u_1, u_2, \ldots, u_n]^T$, $\hat{\beta} = diag\{\hat{\beta}_1, \ldots, \hat{\beta}_n\}$ with $\hat{\beta}_m$ being the estimate of β_i, $m = 1. \ldots, n$, $\hat{\beta}^{-1} = diag\{\frac{1}{\hat{\beta}_1}, \ldots, \frac{1}{\hat{\beta}_n}\}$. Compared with the control designing described previously in (6.10), it is obvious that its denominator contains $\hat{\beta}_i$. If there is no limit on the value of $\hat{\beta}_i$, a singularity problem will occur. Obviously, the singularity problem occurs once the estimate of β is equal to 0, i.e., $\hat{\beta}_i = 0$. Therefore, it is $\frac{1}{\beta_i}$ not β_i that is estimated in this work. Correspondingly, it is $\hat{\beta}_{-1,i}$ not $\frac{1}{\hat{\beta}_i}$ that occurs in (6.10). By designing in this way, the singular value problem is avoided.

Here, the first main result in this chapter is given.

Theorem 6.1 *Consider the leader-following systems (6.1) and (6.2) with gain faults (6.7) under Assumptions 6.1 and 6.2, by constructing the distributed controller (6.10) and adaptive law (6.12), then all followers asymptotically synchronize to the leader with the tracking errors $\delta_i(t)$, $i = 1, \ldots, n$ converging asymptotically to an adjustable neighborhood of the origin, i.e., Ω_1 defined as*

$$\Omega_1 = \left\{ \delta_i \mid |\delta_i| \le \frac{1}{\underline{\varsigma}(B+L)} \sqrt{\frac{\frac{\mu_\beta}{\lambda_\beta} + V_2(0)}{\underline{\varsigma}(P)}}, \quad i = 1, \ldots, n \right\}$$

Proof In the proof, the first task is to prove that the boundedness of $\hat{\beta}_{-1.i}$ is guaranteed by adaptive law (6.12), namely, $\underline{\hat{\beta}}_{-1.i} < \hat{\beta}_{-1.i} < \bar{\hat{\beta}}_{-1.i}$.

Define

$$V_{g\beta,i} = \frac{1}{2} \hat{\beta}_{-1,i}^2$$

Differentiating it with respect to time t, it yields

$$\dot{V}_{g\beta,i} = \hat{\beta}_{-1,i} \dot{\hat{\beta}}_{-1,i}$$

Recalls (6.12), if its first condition holds, then

$$\dot{V}_{g\beta,i} = 0;$$

if its second condition holds, namely,

$$\underline{\hat{\beta}}_{-1.i} < \hat{\beta}_{-1.i} < \bar{\hat{\beta}}_{-1.i} \text{or } \hat{\beta}_{-1.i}\zeta_i < 0 \text{and } (\hat{\beta}_{-1.i} = \underline{\hat{\beta}}_{-1.i} \text{ or } \hat{\beta}_{-1.i} = \bar{\hat{\beta}}_{-1.i}),$$

one has

$$\dot{V}_{g\beta,i} < 0$$

Thus, by using (6.12), $\dot{V}_{g\beta,i} \leq 0$. From the above analysis, one obtains the result: for $\hat{\beta}_{-1,i}$, if its initial value is bounded, (6.12) can guarantee that $\hat{\beta}_{-1,i}$ is also bounded, namely,

$$\underline{\hat{\beta}}_{-1.i} < \hat{\beta}_{-1,i}(t) < \bar{\hat{\beta}}_{-1.i}, \text{ for } t > 0 \text{ if } \underline{\hat{\beta}}_{-1.i} < \hat{\beta}_{-1.i}(0) < \bar{\hat{\beta}}_{-1.i}.$$

Define

$$V_1 = \frac{1}{2}\varepsilon^T P \varepsilon$$

where P is defined as in Lemma 6.2.

Considering (6.5) and (6.10), differentiating V_1 with respect to time t, it yields

$$\dot{V}_1 = \varepsilon^T P \dot{\varepsilon} \quad = -\varepsilon^T P(B + L)(u - \dot{\underline{z}}_0)$$

As stated previously, because z_0 considered in the chapter is assumed to be static, $\dot{z}_0 = 0$. Thus, one has

$$\begin{aligned} \dot{V}_1 &= -\varepsilon^T P(B + L)(u - \dot{\underline{z}}_0) \\ &= -\varepsilon^T P(B + L)u \\ &= -h_1\varepsilon^T P(B + L)(B + D)^{-1}\hat{\beta}_{-1}\varepsilon_a \\ &= -h_1\varepsilon^T P(B + L)(B + D)^{-1}\hat{\beta}_{-1}\beta\varepsilon \end{aligned} \tag{6.13}$$

where $u = [u_1, u_2, \ldots, u_n]^T$, $\hat{\beta}_{-1} = diag\{\hat{\beta}_{-1,1}, \ldots, \hat{\beta}_{-1,n}\}$ and $\beta = diag\{\beta_1, \ldots, \beta_n\}$.

From adaptive law (6.12) and Assumption 6.2, $\beta > 0$ and $\hat{\beta}_{-1} > 0$. Furthermore, it follows from Lemmas 6.1 and 6.2 that $B + L > 0$ and $P > 0$. Hence, from (6.13) and $h_1 > 0$, one has

$$\begin{aligned} \dot{V}_1 &= -h_1\varepsilon^T P(B + L)(B + D)^{-1}\beta\hat{\beta}_{-1}\varepsilon \\ &\leq -\frac{h_1\underline{\beta}\underline{\hat{\beta}}_{-1}\underline{\varsigma}(P)\underline{\varsigma}(B + L)}{\bar{\varsigma}(B + D)}\varepsilon^T\varepsilon \end{aligned} \tag{6.14}$$

where $\underline{\hat{\beta}}_{-1} = \min\{\underline{\hat{\beta}}_{-1,1}, \ldots, \underline{\hat{\beta}}_{-1,n}\}$ and $\underline{\beta} = \min\{\underline{\beta}, \ldots, \underline{\beta}_n\}$.

Now, the term $\tilde{\beta}_{-1}\varepsilon_a^T(D+B)\varepsilon_a$ is added to and subtracted from the right hand of (6.14), then we have

$$\dot{V}_1 \le -\frac{h_1\underline{\beta}\hat{\underline{\beta}}_{-1}\underline{\varsigma}(P)\underline{\varsigma}(B+L)}{\bar{\varsigma}(B+D)}\varepsilon^T\varepsilon + \tilde{\beta}_{-1}\varepsilon_a^T(B+D)\varepsilon_a - \tilde{\beta}_{-1}\varepsilon_a^T(B+D)\varepsilon_a \tag{6.15}$$

where $\tilde{\beta}_{-1} = diag\{\tilde{\beta}_{-1,1}, \ldots, \tilde{\beta}_{-1,n}\}$, and $\tilde{\beta}_{-1,i} = \hat{\beta}_{-1,i} - \beta_{-1,i}, i = 1, \ldots, n$.

From Assumption 6.1, one has, matrix $D + B \ge 0$ is symmetric. And because adaptive law (6.12 and) Assumption 6.2 respectively ensure $\hat{\beta}_{-1,i}$ and $\beta_{-1,i}$ are also bounded, one has

$$\begin{aligned}
&\tilde{\beta}_{-1}\varepsilon_a^T(B+D)\varepsilon_a - \tilde{\beta}_{-1}\varepsilon_a^T(B+D)\varepsilon_a \\
&= (\hat{\beta}_{-1} - \beta_{-1})\varepsilon_a^T(B+D)\varepsilon_a - \tilde{\beta}_{-1}\varepsilon_a^T(B+D)\varepsilon_a \\
&\le (\bar{\hat{\beta}}_{-1} + \bar{\beta}_{-1})\varepsilon_a^T(B+D)\varepsilon_a - \tilde{\beta}_{-1}\varepsilon_a^T(B+D)\varepsilon_a \\
&= (\bar{\hat{\beta}}_{-1} + \bar{\beta}_{-1})\varepsilon^T\beta^T(B+D)\beta\varepsilon - \tilde{\beta}_{-1}\varepsilon_a^T(B+D)\varepsilon_a \\
&\le (\bar{\hat{\beta}}_{-1} + \bar{\beta}_{-1})\bar{\beta}^2\varepsilon^T(B+D)\varepsilon - \tilde{\beta}_{-1}\varepsilon_a^T(B+D)\varepsilon_a
\end{aligned}$$

where $\hat{\beta}_{-1} = diag\{\hat{\beta}_{-1,1}, \ldots, \hat{\beta}_{-1,n}\}$, $\beta_{-1} = diag\{\beta_{-1,1}, \ldots, \beta_{-1,n}\}$, $\bar{\beta} = \max\{\bar{\beta}_1, \ldots, \bar{\beta}_n\}$, $\bar{\beta}_{-1} = \max\{\bar{\beta}_{-1,1}, \ldots, \bar{\beta}_{-1,n}\}$ and $\bar{\hat{\beta}}_{-1} = \max\{\bar{\hat{\beta}}_{-1,1}, \ldots, \bar{\hat{\beta}}_{-1,n}\}$.

Thus, (6.15) can be rewritten as follows:

$$\dot{V}_1 \le -\frac{h_1\underline{\beta}\hat{\underline{\beta}}_{-1}\underline{\varsigma}(P)\underline{\varsigma}(B+L)}{\bar{\varsigma}(B+D)}\varepsilon^T\varepsilon + (\bar{\hat{\beta}}_{-1} + \bar{\beta}_{-1})\bar{\beta}^2\varepsilon^T(B+D)\varepsilon - \tilde{\beta}_{-1}\varepsilon_a^T(B+D)\varepsilon_a \tag{6.16}$$

From (6.11), one further obtains

$$\dot{V}_1 \le -\lambda_1\varepsilon^T\varepsilon - \tilde{\beta}_{-1}\varepsilon_a^T(B+D)\varepsilon_a \tag{6.17}$$

Define

$$V_2 = V_1 + \frac{1}{2\eta_\beta}\tilde{\beta}_{-1}^T\tilde{\beta}_{-1} \tag{6.18}$$

where $\tilde{\beta}_{-1} = \hat{\beta}_{-1} - \beta_{-1}$.

Differentiating V_2 with respect to time t, one has

$$
\begin{aligned}
\dot{V}_2 &= \dot{V}_1 + \frac{1}{\eta_\beta}\tilde{\beta}_{-1}^T\dot{\hat{\beta}}_{-1} - \frac{1}{\eta_\beta}\tilde{\beta}_{-1}^T\dot{\beta}_{-1} \\
&\le -\frac{h_1\underline{\beta}\underline{\hat{\beta}}_{-1}\underline{\varsigma}(P)\underline{\varsigma}(B+L)}{\bar{\varsigma}(B+D)}\varepsilon^T\varepsilon + \\
&\quad (\bar{\hat{\beta}}_{-1} + \bar{\beta}_{-1})\bar{\beta}^2\varepsilon^T(B+D)\varepsilon - \tilde{\beta}_{-1}\varepsilon_a^T(B+D)\varepsilon_a + \\
&\quad \frac{1}{\eta_\beta}\tilde{\beta}_{-1}^T\dot{\hat{\beta}}_{-1} - \frac{1}{\eta_\beta}\tilde{\beta}_{-1}^T\dot{\beta}_{-1}
\end{aligned} \tag{6.19}
$$

where $\dot{\beta}_{-1} = diag\{\dot{\beta}_{-1,1}, \ldots, \dot{\beta}_{-1,n}\}$ and $\dot{\hat{\beta}}_{-1} = diag\{\dot{\hat{\beta}}_{-1,1}, \ldots, \dot{\hat{\beta}}_{-1n}\}$.

It is worth pointing out that if the first condition in (6.12) is true, the corresponding value of the adaptive law is "0", which is also described as

$$0 = \zeta_i - \zeta_i$$

Thus, (6.12) is also re-written as:

$$
\dot{\hat{\beta}}_{-1,i} = \begin{cases} \zeta_i - \zeta_i, \text{ if } (\hat{\beta}_{-1.i} = \underline{\hat{\beta}}_{-1.i} \text{ or } \hat{\beta}_{-1.i} = \bar{\hat{\beta}}_{-1.i}) \\ \quad \text{and } \hat{\beta}_{-1.i}\zeta_i \ge 0 \\ \zeta_i, \text{ if } \underline{\hat{\beta}}_{-1.i} < \hat{\beta}_{-1.i} < \bar{\hat{\beta}}_{-1.i} \text{or } \hat{\beta}_{-1.i}\zeta_i < 0 \\ \quad \text{and } (\hat{\beta}_{-1.i} = \underline{\hat{\beta}}_{-1.i} \text{ or } \hat{\beta}_{-1.i} = \bar{\hat{\beta}}_{-1.i}) \end{cases},
$$

$$\underline{\hat{\beta}}_{-1.i} < \hat{\beta}_{-1.i}(0) < \bar{\hat{\beta}}_{-1.i}$$

Considering the above adaptive law, (6.19) is developed as

$$\dot{V}_2 \le -\lambda_1\varepsilon^T\varepsilon - \frac{\gamma_\beta\tilde{\beta}_{-1}^T\hat{\beta}_{-1}}{\eta_\beta} - \frac{\tilde{\beta}_{-1}^T\dot{\beta}_{-1}}{\eta_\beta} - \sum_{i=1}^{n}\frac{I_i\tilde{\beta}_{-1,i}\zeta_i}{\eta_\beta}$$

where $I_i = 1(or\ 0)$ if the first (second) condition is satisfied.

Notice that, when the first condition of (6.12) is true, one has

$$(\hat{\beta}_{-1.i} = \underline{\hat{\beta}}_{-1.i} \text{ or } \hat{\beta}_{-1.i} = \bar{\hat{\beta}}_{-1.i}) \text{ and } \hat{\beta}_{-1.i}\zeta_i \ge 0$$

and

$$\tilde{\beta}_{-1,i}\hat{\beta}_{-1,i} = \frac{1}{2}(\hat{\beta}_{-1,i}^2 + \tilde{\beta}_{-1,i}^2 - \beta_{-1,i}^2) \quad \ge 0$$

And since

$$\tilde{\beta}_{-1,i}\zeta_i = \frac{1}{\hat{\beta}_{-1,i}^2}\tilde{\beta}_{-1,i}\hat{\beta}_{-1,i}\hat{\beta}_{-1,i}\zeta_i \ge 0$$

we have

$$-\sum_{i=1}^{n}\frac{I_i\tilde{\beta}_{-1,i}\zeta_i}{\eta_\beta}\le 0$$

Further, one has

$$\begin{aligned}\dot{V}_2 &\le -\lambda_1 e^T e-\frac{\gamma_\beta}{\eta_\beta}\tilde{\beta}_{-1}^T\hat{\beta}_{-1}-\frac{1}{\eta_\beta}\tilde{\beta}_{-1}^T\dot{\beta}_{-1}\\ &\le -\lambda_1\varepsilon^T\varepsilon-\frac{\gamma_\beta}{\eta_\beta}\tilde{\beta}_{-1}^T\tilde{\beta}_{-1}+\frac{\gamma_\beta}{\eta_\beta}\beta_{-1}^T\beta_{-1}-\frac{1}{\eta_\beta}\tilde{\beta}_{-1}^T\dot{\beta}\\ &\le -\lambda_1\varepsilon^T\varepsilon-\frac{\gamma_\beta}{\eta_\beta}\tilde{\beta}_{-1}^T\tilde{\beta}_{-1}+n\left(\frac{\gamma_\beta}{\eta_\beta}\bar{\beta}^2+\frac{2}{\eta_\beta}\bar{\beta}\bar{\bar{\beta}}\right)\\ &\le -\lambda_1\varepsilon^T\varepsilon-\frac{\gamma_\beta}{\eta_\beta}\tilde{\beta}_{-1}^T\tilde{\beta}_{-1}+\mu_\beta\\ &\le -\lambda_\beta V_2+\mu_\beta\end{aligned}\tag{6.20}$$

where $\mu_\beta=n(\frac{\gamma_\beta}{\eta_\beta}\bar{\beta}^2+\frac{2}{\eta_\beta}\bar{\beta}\bar{\bar{\beta}})$ and $\lambda_\beta=\min\{\frac{\lambda_1}{\bar{\varsigma}(P)\bar{\varsigma}(B+L)},\frac{\gamma_\beta}{\eta_\beta}\}$.

From (6.20), one has

$$\begin{aligned}V_2&\le\frac{\mu_\beta}{\lambda_\beta}+[V_2(0)-\frac{\mu_\beta}{\lambda_\beta}]e^{-\lambda_\beta t}\\ &\le\frac{\mu_\beta}{\lambda_\beta}+V_2(0)\end{aligned}\tag{6.21}$$

Because of $V_2=\frac{1}{2}\varepsilon^T P\varepsilon+\frac{1}{2\eta_\beta}\tilde{\beta}_{-1}^T\tilde{\beta}_{-1}$, one has

$$||\varepsilon||\le\sqrt{\frac{2(\frac{\mu_\beta}{\lambda_\beta}+V_2(0))}{\underline{\varsigma}(P)}},\quad |\tilde{\beta}_{-1,i}|\le\sqrt{2\eta_\beta\left(\frac{\mu_\beta}{\lambda_\beta}+V_2(0)\right)}$$

Based on Lemma 6.3, one has

$$||\delta||\le\frac{||\varepsilon||}{\underline{\varsigma}(L+B)}\le\frac{\sqrt{\frac{2(\frac{\mu_\beta}{\lambda_\beta}+V_2(0))}{\underline{\varsigma}(P)}}}{\underline{\varsigma}(L+B)}$$

Because of $\delta=[\delta_1,\ldots,\delta_n]^T$, one has

$$|\delta_i|\le\frac{||\varepsilon||}{\underline{\varsigma}(L+B)}\le\frac{\sqrt{\frac{2(\frac{\mu_\beta}{\lambda_\beta}+V_2(0))}{\underline{\varsigma}(P)}}}{\underline{\varsigma}(L+B)},\quad i=1,\ldots,n$$

which means that each follower in the graph synchronizes asymptotically to the leader. The proof is completed here.

Remark 6.6 If the gain communication Fault β is not taken into account, the control law (6.10) becomes the following form:

$$u_i = \frac{h_1}{(d_i + b_i)} \varepsilon_{a,i}.$$

considering the above control law, (6.13) becomes

$$\begin{aligned}
\dot{V}_1 &= -\varepsilon^T P(B+L)(u - \dot{\underline{z}}_0) \\
&= -\varepsilon^T P(B+L)u \\
&= -h_1 \varepsilon^T P(B+L)(B+D)^{-1}\varepsilon_a \\
&= -h_1 \beta \varepsilon^T P(B+L)(B+D)^{-1}\varepsilon
\end{aligned}$$

From Assumptions 6.1 and 6.2, and Lemma 6.1, we know, $h_1 > 0, \beta > 0$ (Assumption 6.2), $P > 0$ and $B + L > 0$ (Assumption 6.1 and Lemma 6.1), $P(L+B)(D+B)^{-1} > 0$. Thus, $\dot{V}_1 < 0$ when $\varepsilon \neq 0$. From Lyapunov stability theory, it can be obtained that $\lim_{t \to +\infty} \varepsilon(t) = 0$, which means that the synchronization between the leader and followers has been obtained. That is, for the communication gain fault, the control law of each follower designed in the ideal case (without the communication fault) also guarantees the consensus even in the case with the communication fault. However, it causes a weaker consensus performance, and the convergence time will become larger, which will be shown in Figs. 6.2 and 6.3 in Simulation.

6.3.2 *Fault-Tolerant Consensus Control Against Transmission Bias Faults*

In this subsection, let us consider the communication bias fault described in (6.8).

From the communication fault model (6.8), one has $\varepsilon_{a,i} = \varepsilon_i + w_i$. Further, one has

$$\varepsilon_i = \varepsilon_{a,i} - w_i.$$

To achieve the synchronization between the leader and followers, the following adaptive control law is constructed for each follower in the graph:

$$u_i = h_2(\varepsilon_{a,i} - \hat{w}_i) \tag{6.22}$$

where $\hat{w}_i$ is the estimate of w_i, $h_2 \in R$ is design parameter.

Remark 6.7 Different from (6.10), $\hat{w}_i$ is added in (6.22). In fact, $\hat{w}_i$ is a compensation term of the communication bias fault w_i, whose purpose is to reduce the adverse influence of the fault w_i.

Define the adaptive law of $\hat{w}_i$ $(i = 1, \ldots, n)$

$$\dot{\hat{w}}_i = -\eta_w(\sum\nolimits_{i=1}^{n} h_2(d_i + b_i)(\varepsilon_{a,i} + \hat{w}_i)) - \kappa_i \hat{w}_i \tag{6.23}$$

where adaptive rate $\eta_w > 0 \in R$, $\kappa_i > 0$ denotes a design constant, which satisfies

$$\frac{\kappa_i}{2\eta_w} - \sum\nolimits_{i=1}^{n} (r + h_2 \bar{\varsigma}(D + B)) > \gamma_w \tag{6.24}$$

with $\gamma_w > 0$ are design parameters, and $r > 0$ represents a real constant given in (6.28).

Now, let us give another main result in this chapter.

Theorem 6.2 *Consider the distributed system (6.1) and leader node (6.2) with bias faults (6.8) under Assumptions 6.1 and 6.2, by constructing the distributed controller (6.22) and the adaptive law (6.23), then all followers in the graph synchronize asymptotically to the leader node 0 with the tracking errors $\delta_i(t)$ asymptotically converging to an adjustable neighborhood of the origin, i.e., Ω_2 defined as*

$$\Omega_2 = \left\{ \delta_i \middle| \, |\delta_i| \leq \frac{1}{\underline{\varsigma}(B + L)} \sqrt{\frac{\frac{\mu_w}{\lambda_w} + V_4(0)}{\underline{\varsigma}(P)\underline{\varsigma}(B + L)}}, \quad i = 1, \ldots, n \right\}$$

Proof Define

$$V_3 = \frac{1}{2}\varepsilon^T \varepsilon$$

Differentiating V_3 with respect to time t and considering (6.22), it yields

$$\begin{aligned} \dot{V}_3 &= \varepsilon^T \dot{\varepsilon} = -e^T (B + L)(u - \dot{\underline{z}}_0) \\ &= -\varepsilon^T (B + L)u \\ &= -h_2 \varepsilon^T (B + L)(\varepsilon_a - h_2 \hat{w}_v) \\ &= -h_2 \varepsilon^T (B + L)e + h_2 \varepsilon^T (B + L)\tilde{w}_v \\ &= -h_2 \varepsilon^T (B + L)\varepsilon + \\ &\quad h_2 \varepsilon^T (B + D)\tilde{w}_v - h_2 \varepsilon^T A \tilde{w}_v \end{aligned} \tag{6.25}$$

where $u = [u_1, u_2, \ldots, u_n]^T$, $\varepsilon_a = [\varepsilon_{a1}, \ldots, \varepsilon_{an}]^T$, $\tilde{w}_v = \hat{w}_v - w_v = [\tilde{w}_1, \ldots, \tilde{w}_n]^T$, $\hat{w}_v = [\hat{w}_1, \ldots, \hat{w}_n]^T$, $\hat{w}_i$ is the estimate of w_i, $w_v = [w_i, \ldots, w_n]^T$, $\tilde{w}_i = \hat{w}_i - w_i$.

Since

$$
\begin{aligned}
& h_2\varepsilon^T(B+D)\tilde{w}_v \\
&\quad = h_2\varepsilon_a^T(B+D)\tilde{w}_v - h_2 w_v^T(B+D)\tilde{w}_v \\
&\quad = h_2\varepsilon_a^T(B+D)\tilde{w}_v \\
&\qquad - h_2\hat{w}_v^T(B+D)\tilde{w}_v + h_2\tilde{w}_v^T(B+D)\tilde{w}_v
\end{aligned}
\tag{6.26}
$$

and

$$
-h_2\varepsilon^T A\tilde{w}_v \le r h_2\tilde{w}_v^T\tilde{w}_v + \frac{1}{r}h_2\varepsilon^T AA^T\varepsilon \tag{6.27}
$$

where design parameter $r > 0$, which satisfies

$$
-\frac{1}{r}\bar{\varsigma}(AA^T) + \underline{\varsigma}(B+L) > \lambda_2 \tag{6.28}
$$

where design parameter $\lambda_2 > 0$, then (6.25) is developed as

$$
\begin{aligned}
\dot{V}_3 \le & -h_2\varepsilon^T(B+L)\varepsilon + h_2\varepsilon_a^T(B+D)\tilde{w}_v - \\
& h_2\hat{w}_v^T(B+D)\tilde{w}_v + h_2\tilde{w}_v^T(B+D)\tilde{w}_v + \\
& r h_2\tilde{w}_v^T\tilde{w}_v + \frac{1}{r}h_2 e AA^T\varepsilon \\
\le & -h_2\left(\underline{\varsigma}(B+L) - \frac{\bar{\varsigma}(AA^T)}{r}\right)\varepsilon^T\varepsilon + \\
& h_2\tilde{w}_v^T(B+D)^T(\varepsilon_a + \hat{w}_v) + \\
& h_2(r + \bar{\varsigma}(B+D))\tilde{w}_v^T\tilde{w}_v \\
\le & -h_2\left(\underline{\varsigma}(B+L) - \frac{\bar{\varsigma}(AA^T)}{r}\right)\varepsilon^T\varepsilon + \\
& \sum\nolimits_{i=1}^{n} h_2(d_i + b_i)(\varepsilon_{a,i} + \hat{w}_i)\tilde{w}_i + \\
& \sum\nolimits_{i=1}^{n} h_2(r + \bar{\varsigma}(B+D))\tilde{w}_i^2
\end{aligned}
\tag{6.29}
$$

Define

$$
V_4 = V_3 + \frac{1}{\eta_w}\tilde{w}_v^T\tilde{w}_v
$$

where $\eta_w > 0$. Differentiating V_4 with respect to time t and considering (6.29), it gives

$$\begin{aligned}\dot{V}_4 \leq &-h_2\left(\underline{\varsigma}(L+B)-\frac{\bar{\varsigma}(AA^T)}{r}\right)\varepsilon^T\varepsilon+\\&\sum\nolimits_{i=1}^{n}h_2(b_i+d_i)(\varepsilon_{a,i}+\hat{w}_i)\tilde{w}_i+\\&\sum\nolimits_{i=1}^{n}(r+h_2\bar{\varsigma}(D+B))\tilde{w}_i^2+\\&\frac{1}{\eta_w}\tilde{w}_v^T\dot{\hat{w}}_v-\frac{1}{\eta_w}\tilde{w}_v^T\dot{w}_v\end{aligned} \tag{6.30}$$

Substituting (6.23) into (6.30), from (6.24) and (6.28), it yields

$$\begin{aligned}\dot{V}_4 \leq &-h_2\lambda_2\varepsilon^T\varepsilon-\frac{1}{\eta_w}\tilde{w}_v^T\dot{w}_v-\sum\nolimits_{i=1}^{n}\frac{\kappa_i}{\eta_w}\tilde{w}_i\hat{w}_i+\\&\sum\nolimits_{i=1}^{n}(r+h_2\bar{\varsigma}(D+B))\tilde{w}_i^2\end{aligned}$$

Since

$$\begin{aligned}\frac{\kappa_i}{\eta_w}\tilde{w}_i\hat{w}_i &= \frac{\kappa_i}{\eta_w}[\tilde{w}_i(\hat{w}_i-w_i)]\\&=\frac{\kappa_i}{\eta_w}(-\tilde{w}_i^2+\tilde{w}_iw_i)\\&\leq\frac{\kappa_i}{\eta_w}(-\tilde{w}_i^2+\frac{1}{2}\tilde{w}_i^2+\frac{1}{2}w_i^2)\\&=-\frac{\kappa_i}{2\eta_w}\tilde{w}_i^2+\frac{\kappa_i}{2\eta_w}w_i^2\end{aligned}$$

we have

$$\begin{aligned}\dot{V}_4 \leq &-h_2\lambda_2\varepsilon^T\varepsilon-\frac{1}{\eta_w}\tilde{w}_v^T\dot{w}_v-\sum\nolimits_{i=1}^{n}\frac{\kappa_i}{2\eta_w}\tilde{w}_i^2+\\&\sum\nolimits_{i=1}^{n}\frac{\kappa_i}{2\eta_w}w_i^2+\sum\nolimits_{i=1}^{n}(r+h_2\bar{\varsigma}(D+B))\tilde{w}_i^2\end{aligned}$$

From (6.24), it follows that

$$\begin{aligned}\dot{V}_4 \leq &-h_2\lambda_2\varepsilon^T\varepsilon-\sum\nolimits_{i=1}^{n}\gamma_w\tilde{w}_i^2+\\&\sum\nolimits_{i=1}^{n}\frac{\kappa_i}{2\eta_w}w_i^2-\frac{1}{\eta_w}\tilde{w}_v^T\dot{w}_v\\&\leq -h_2\lambda_2h_2\varepsilon^T\varepsilon+\mu_w-\gamma_w\tilde{w}_v^T\tilde{w}_v\\&\leq\mu_w-\lambda_wV_4\end{aligned} \tag{6.31}$$

where

$$\mu_w=\frac{\kappa}{\eta_w}n\bar{w}+\frac{1}{\eta_w}n\bar{w}\bar{\bar{w}},\ \lambda_w=\min\{\frac{h_2\lambda_2}{\bar{\varsigma}(L+B)},\frac{\gamma_w}{\eta_w}\}$$

with

$$\bar{w} = \max\{\bar{w}_1, \ldots, \bar{w}_n\}, \kappa = \max\{\kappa_1, \ldots, \kappa_n\}, \ \bar{\bar{w}} = \max\{\bar{\bar{w}}_1, \ldots, \bar{\bar{w}}_n\}.$$

From (6.31), one further has

$$\begin{aligned} V_4 &\le \frac{\mu_w}{\lambda_w} + [V_4(0) - \frac{\mu_w}{\lambda_w}]e^{-\lambda_w t} \\ &\le \frac{\mu_w}{\lambda_w} + V_4(0) \end{aligned} \tag{6.32}$$

Since $V_4 = \frac{1}{2}\varepsilon^T\varepsilon + \frac{1}{2\eta_w}\tilde{w}_v^T\tilde{w}_v$, one has

$$||\varepsilon|| \le \sqrt{2(\frac{\mu_w}{\lambda_w} + V_4(0))}, \ \ ||\tilde{w}_v|| \le \sqrt{2\eta_w(\frac{\mu_w}{\lambda_w} + V_4(0))}$$

From Lemma 6.3, it follows

$$||\delta|| \le \frac{||\varepsilon||}{\underline{\varsigma}(B+L)} \le \frac{\sqrt{2(\frac{\mu_w}{\lambda_w} + V_4(0))}}{\underline{\varsigma}(B+L)}$$

Because of $\delta = [\delta_1, \ldots, \delta_n]^T$, one has

$$|\delta_i| \le \frac{1}{\underline{\varsigma}(B+L)}\sqrt{2\left(\frac{\mu_w}{\lambda_w} + V_4(0)\right)}, i = 1, \ldots, n$$

which indicates that all followers asymptotically synchronize to the leader, and the proof is completed.

Remark 6.8 Unlike the communication gain faults investigated in the previous subsection, once the communication bias faults are not compensated, the control input (6.22) becomes $u = h_2\varepsilon_a$. Substituting it into $\dot{V}_3$, it yields

$$\begin{aligned} \dot{V}_3 - \varepsilon^T\dot{\varepsilon} &= -\varepsilon^T(B+L)u \\ &= -h_2\varepsilon^T(B+L)\varepsilon_a \\ &= -h_2\varepsilon^T(B+L)\varepsilon + h_2\varepsilon^T(B+L)w_v \end{aligned}$$

Obviously, because there exists an additional term $h_2\varepsilon^T(L+B)w_v$, it cannot guarantee that the closed-loop system is stable. Therefore, in order to obtain the synchronization between the followers and leader, the communication bias fault w must be compensated as doing (6.22).

6.4 Simulation Results

Consider a digraph $\mathcal{G}$ containing five followers and a leader described in Fig. 6.1, where the weights on edges are equal to one. The followers have the same form as (6.1). In this work, for the leader, its dynamics is assumed to be $\dot{z}_0 = 0$ and $z_0 = 0$.

The communication fault models (6.7) and (6.8) have the following forms:

$$\varepsilon_{a,i} = \beta_i \varepsilon_i, \quad \varepsilon_{a,i} = \varepsilon_i + w_i.$$

In this work, the fault parameters are assumed that $\beta_1 = 0.5$, $\beta_2 = \cdots = \beta_5 = 0$ and $w_1 = cos(t + pi * 3.9/5) * 0.1$, and $w_2 = \cdots = w_5 = 0$.

Design parameters are set as: $h_1 = h_2 = 30$, $\eta_\beta = \eta_w = 5$, $\gamma_\beta = \gamma_w = 0.01$, $\nu = 10$, $r = 10$, $\lambda_1 = \lambda_2 = 0.01$, and $\kappa = 20$.

In this simulation, it is assumed that gain and bias faults occur only in a communication link, via which follower 1 obtain the information from the other nodes and all the other links are fault-free. This simulation is developed in the following three cases.

Case 1: All communication links in the graph are fault-free. Under the controllers defined in (6.6), the consensus has been obtained, shown in Fig. 6.2. From Fig. 6.2, it is easy to find that the closed-loop system has good consensus performance.

Case 2: Transmission gain fault occurs in the link, via which follower 1 obtain the information from the other nodes. In this case, the communication gain fault β_4 is set as $\beta = 0.5$. Note that the actual value of the information from the other nodes cannot be obtained by follower 1 because of the communication gain fault β_4. If it is not taken into account in control designing, the state trajectories are shown in Fig. 6.3. From Fig. 6.2, we can find that, although follower 4 is subject to the communication, the consensus also can be obtained under the controllers defined in (6.6). Comparing Fig. 6.2 with Fig. 6.3, however, we also find, although the consensus is obtained, the convergence time of the follower 4 becomes longer. If the fault-tolerant control method defined (6.10) is utilized, the fault is compensated and good performance is obtained, which is shown in Fig. 6.4. Figure 6.5 shows the gain fault and its estimate.

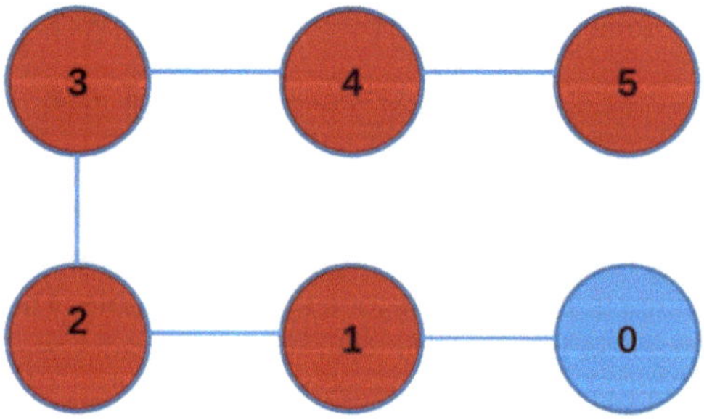

Fig. 6.1 Topology

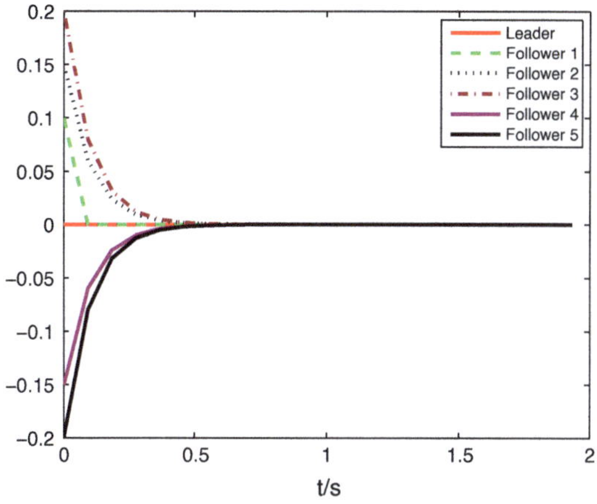

Fig. 6.2 The states in the fault-free case

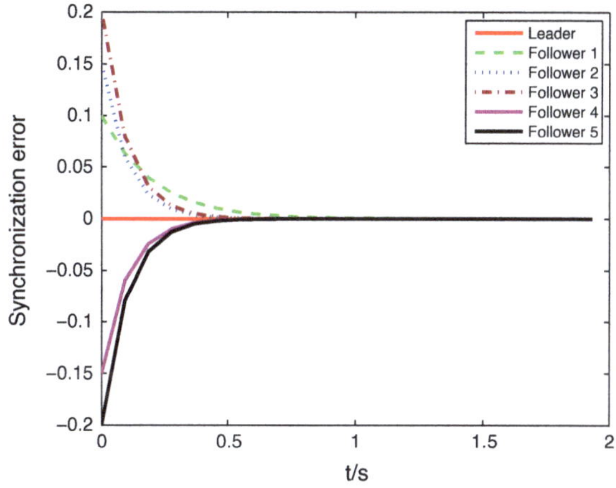

Fig. 6.3 The states in the faulty case controlled by (6.6)

Case 3: Transmission bias fault occurs in the link, via which follower 1 obtain the information from the other nodes. The simulation results in this case are shown in Figs. 6.6 and 6.7. Under fault-tolerant controllers in (6.22), the consensus between the leader and followers is obtained, and the trajectories of the nodes are shown in Fig. 6.6. Figure 6.7 gives the time profiles of the fault and its estimate.

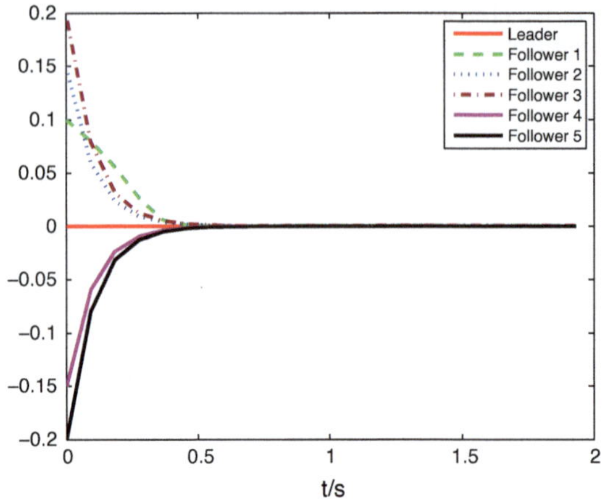

Fig. 6.4 The states in the faulty case controlled by FTC (6.10)

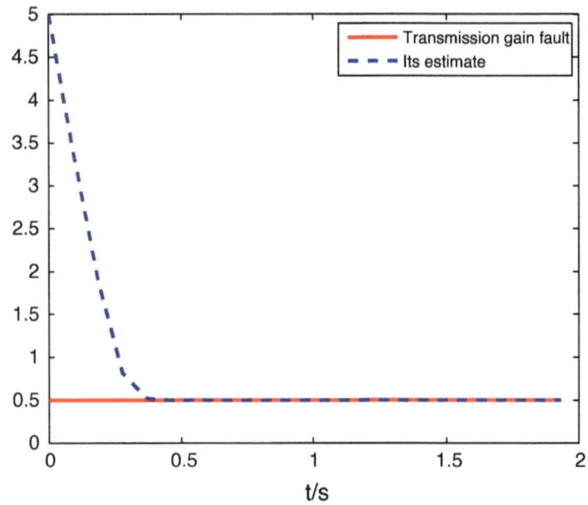

Fig. 6.5 The gain fault and its estimate

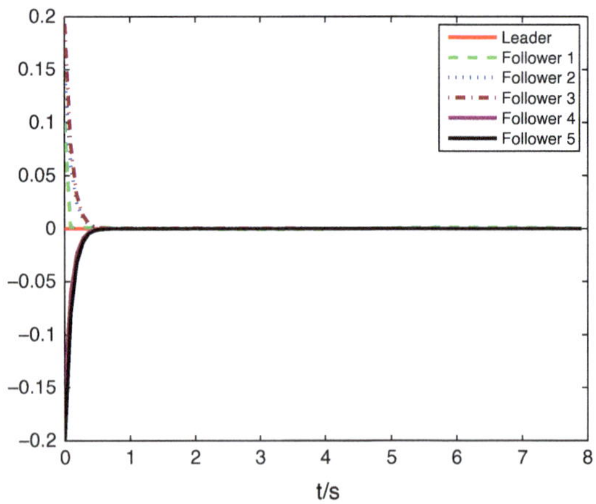

Fig. 6.6 The states in the faulty case controlled by FTC (6.22)

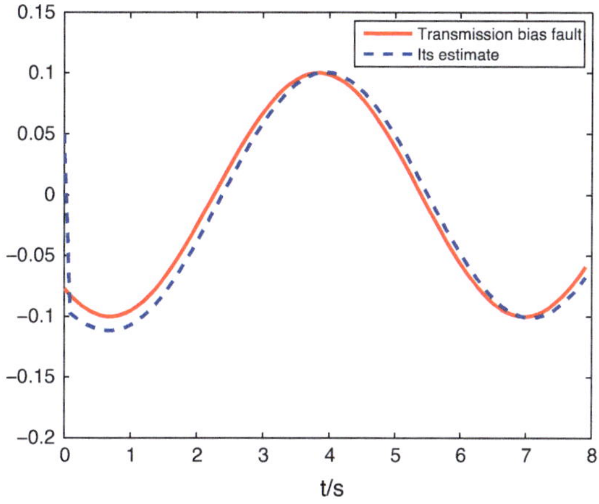

Fig. 6.7 The bias fault and its estimate

6.5 Conclusion

This chapter investigates the leader-following synchronization control problem. For the communication gain and bias faults that occurred in the communication network, two fault-tolerant adaptive control methods are constructed, respectively, to ensure that the synchronization between the leader and followers is achieved and synchronization errors asymptotically converge to an adjustable neighborhood of the origin. Based on algebraic graph theory and Lyapunov theory, the closed-loop system stability analysis is conducted. Simulation results are given to demonstrate the effectiveness and advantages of the new design techniques developed.

References

Chen J, Cao X, Cheng P, Xiao Y, Sun Y (2010) Distributed collaborative control for industrial automation with wireless sensor and actuator networks. IEEE Trans Ind Electron 57(12):4219–4230

Chen Y, Lü J, Lin Z (2013) Consensus of discrete-time multi-agent systems with transmission nonlinearity. Automatica 49:1768–1775

Du H, Li S, Shi P (2012) Robust consensus algorithm for second-order multi-agent systems with external disturbances. Int J Control 85(12):1913–1928

Hong Y, Chen G, Bushnell L (2008) Distributed observers design for leader-following control of multi-agent networks. Automatica 44(3):846–850

Hu J, Feng G (2010) Distributed tracking control of leader-follower multi-agent systems under noisy measurement. Automatica 46:1382–13879

Huang M, Manton JH (2009) Coordination and consensus of networked agents with noisy measurement: stochastic algorithms and asymptotic behavior. SIAM J Control Optim 48(1):134–161

Li T, Zhang JF (2009) Mean square average consensus under measurement noises and fixed topologies: necessary and sufficient conditions. Automatica 45(8):1929–1936

Li ZK, Duan ZS, Chen GR, Huang L (2010) Consensus of multiagent systems and synchronization of complex networks: a unified Viewpoint. IEEE Trans Circuits Syst-I: Regul Pap 57(1):213–224

Murray R (2007) Recent research in cooperative control of multivehicle systems. ASME J Dyn Syst Meas Control 29(5):571–583

Ni W, Cheng D (2010) Leader-following consensus of multi-agent systems under fixed and switching topologies. Syst Control Lett 59:209–217

Peng ZH, Wang D, Zhang HW, Sun G, Wang H (2013) Distributed model reference adaptive control for cooperative tracking of uncertain dynamical multiagent systems. IET Control Theory Appl 7(8):1079–1087

Ren W, Moore K, Chen Y (2007) High-order and model reference consensus algorithms in cooperative control of multivehicle systems. ASME J Dyn Syst Meas Control 129(5):678–688

Ren W, Moore KL, Chen Y (2007) High-order and model reference consensus algorithms in cooperative control of multivehicle systems. ASME J Dyn Syst Meas Control 129(5):678–688

Ren W, Beard R, Atkins E (2007) Information consensus in multivehicle cooperative control. IEEE Control Syst Mag 27(2):71–82

Shen QK, Shi P (2015) Distributed command filtered backstepping consensus tracking control of nonlinear multiple-agent systems in strict-feedback form. Automatica 53:120–124

Shen QK, Jiang B, Shi P (2014) Cooperative adaptive fuzzy tracking control for networked unknown nonlinear multi-agent systems with time-varying actuator faults. IEEE Trans Fuzzy Syst 22(3):494–504

Shi P, Shen Q (2015) K: Cooperative control of multi-agent Systems with unknown state-dependent controlling effects. IEEE Trans Autom Sci Eng 12(3):827–834

Wang X, Chen G (2002) Pinning control of scale-free dynamical networks. Phys A 310(3–4):521–531

Wang W, Slotine J (2006) A theoretical study of different leader roles in networks. IEEE Trans Autom Control 51(7):1156–1161

Wang X, Yadav V, Balakrishnan S (2007) Cooperative UAV formation flying with obstacle/collision avoidance. IEEE Trans Control Syst Technol 15(4):672–679

Yu J, Wang L (2010) Group consensus in multi-agent systems with switching topologies and communication delays. Syst Control Lett 59:340–348

Yu H, Xia X (2012) Adaptive consensus of multi-agents in networks with jointly connected topologies. Automatica 48:1783–1790

Zhang H, Lewis FL (2012) Adaptive cooperative tracking control of higher-order nonlinear systems with unknown dynamics. Automatica 48(7):1432–1439

Zhang H, Lewis FL, Das A (2011) Optimal design for synchronization of cooperative systems: state feedback, observer and output feedback. IEEE Trans Autom Control 56(8):1948–1952

Zheng X, Shen QK (2023) Adaptive fault-tolerant control for high-order nonlinear systems with supervisory controllers and command filters. Int J Control Autom Syst 21(1):12–19

Zhu W, Cheng D (2010) Leader-following consensus of second-order agents with multiple time-varying delays. Automatica 46(12):1994–1999

Chapter 7
Cooperative Fault-Tolerant Neural Control Against Signal Communication Faults

7.1 Introduction

For multi-agent systems (MASs), in the past two decades, distributed control methods and technologies have attracted extensive attention from engineering and theoretical fields, because of their greater efficiency and operational capability Ren and Cao (2011). Up to now, for MASs including leader-following systems (LFSs) considered in this chapter, there are a lot of results reported in literature Bertsekas and Tsitsiklis (2007), Gong et al. (2019), Moreau (2005), Ren et al. (2007), Shen et al. (2020), Zhang and Lewis (2012), Shi and Shen (2017), Shen and Shi (2016), Zhao et al. (2018), Jadbabaie et al. (2003). For cooperative control of MASs, there are many research directions, for example, cooperative regulator and cooperative tracking problems. Among the directions, consensus or synchronization control problem is fundamental but important Shen et al. (2020). For a class of single integrator or first-order MASs or LFSs, their consensus control was studied in Bertsekas and Tsitsiklis (2007), Gong et al. (2019). In Moreau (2005), for discrete MASs on balanced digraphs, consensus problem was considered, while Ren et al. (2007) considered the case where the graph has the weighted directed communication topology and derived the corresponding sufficient condition. In Zhang and Lewis (2012) and Shi and Shen (2017), the synchronization, tracking problems were studied, respectively. Considering the case where the followers were subject dead-zone, Shen and Shi (2016) investigated the consensus tracking problems of LFSs. In the above results, however, system faults especially sensor faults are not taken into account. In practice, the components of systems and communication networks such as actuator, sensors, and communication links often fail. For complex systems such as MASs and LFSs, how to decrease the influence of the faults is important.

Generally speaking, in practical application, system components such as actuators and sensors may become faulty. What's more, These failures will adversely affect the controlled systems. Therefore, lots of effective fault-tolerant control (FTC) methods have been proposed such that the controlled systems' reliability can be improved

Q. Shen, *Distributed Fault-Tolerant Consensus Control of Leader-Following Systems*, Intelligent Control and Learning Systems 11,
https://doi.org/10.1007/978-981-99-7426-9_7

and the stability of the controlled system can be guaranteed in all situations Shen et al. (2021), Zhu et al. (2023), Xu and Shen (2022), Semsar and Khorasani (2008), Zheng and Shen (2023), Deng and Yang (2017), Ye et al. (2019), Ye et al. (2016), Deng and Yang (2016), Hua et al. (2018), Deng and Yang (2018). In recent years, for MASs including LFSs, actuator faults, transmission nonlinearities or faults in communication links were, respectively, investigated in Shen et al. (2014), Shen et al. (2019), Ma and Ma (2018). However, only actuator faults were concerned in Shen et al. (2014), Ma and Ma (2018), Ma and Yang (2016), Jin et al. (2018), Zuo et al. (2015), Deng and Yang (2017), Ye et al. (2019), Ye et al. (2016), Deng and Yang (2016), Hua et al. (2018), Deng and Yang (2018).

In this chapter, the distributed adaptive fault-tolerant tracking control problem is considered for a class of uncertain nonlinear leader-following systems, an FTC scheme against signal communication faults is designed. Compared with the results in Gong et al. (2019), Moreau (2005), Ren et al. (2007), Shen et al. (2020), Zhang and Lewis (2012), Shi and Shen (2017), Shen and Shi (2016), Zhao et al. (2018), signal communication faults are investigated, and the corresponding compensation algorithm is designed for the faults.

The rest of the chapter is organized as follows. In Sect. 7.2, basic graph theory is introduced, meanwhile radial basis function (RBF) neural networks (NNs) are described. The problem considered in this chapter is also formulated in this section. In Sect. 7.3, the main results obtained in this chapter are given. Section 7.4 gives some simulation results. Finally, the conclusion is drawn in Sect. 7.5.

7.2 Preliminaries and Problem Formation

7.2.1 Basic Graph Theory

As stated in Chap. 3, in the sense of mathematics, MASs including LFSs investigated in this book can be represented by a graph. Obviously, it involves basic graph theory in the research of MASs. So, let us first introduce it in this subsection. For basic graph theory, it was already introduced in the previous chapter. For the convenience of reading, its related knowledge will be re-introduced here again. If the readers have gained an understanding of the basic graph theory, the readers can skip this part and read the following content directly.

A class of multi-agent systems will be considered here, which consists of N followers and a leader. For the leader and followers, the information exchange among the leader and followers is often modeled as a graph Semsar and Khorasani (2008). Hence, we first introduce the knowledge of basic graph theory in this subsection.

A directed graph can be represented as $\mathscr{G} = (\mathscr{V}, \mathscr{E})$, where agent set $\mathscr{V}$ and edge (arcs) set $\mathscr{E}$ are defined as $\mathscr{V} = (\mathscr{V}_1, \ldots, \mathscr{V}_N)$ and $\mathscr{E} \subseteq \mathscr{V} \times \mathscr{V}$, respectively. Note that, if there exists an edge from node i to node j, then $(\mathscr{V}_i, \mathscr{V}_j) \in E$. For $\mathscr{G}$, its adjacency matrix $\mathscr{A}$ can be described as $\mathscr{A} = [a_{ij}] \in R^{N \times N}$, where element a_{ij} is defined as follows: (1) $a_{ij} = 0$ if there exists no an edge from node i to node j, i.e.,

$(\mathscr{V}_i, \mathscr{V}_j) \notin \mathscr{E}$; and (2) $a_{ij} \neq 0$ if there exists no an edge from node i to node j, i.e., $(\mathscr{V}_i, \mathscr{V}_j) \in E$. Throughout this chapter, a_{ii} is assumed to be equal to zero, i.e.,$a_{ii} = 0$, and the topology is also assumed to be fixed, which means that $\mathscr{A}$ is fixed and is not variant.

In directed graph $\mathscr{G}$, for the ith agent, its in-degree d_i is defined as $d_i = \sum_{j=1}^{N} a_{ij}$. Correspondingly, for the directed graph $\mathscr{G}$, its in-degree matrix and Laplacian matrix, i.e., $\mathscr{D}$ and $\mathscr{L}$, are also defined respectively as $\mathscr{D} = diag\{d_1, \ldots, d_N\} \in R^{N \times N}$ and $\mathscr{L} = [l_{ij}] = \mathscr{D} - \mathscr{A} \in R^{N \times N}$. For the Laplacian matrix $\mathscr{L}$, $L\underline{1} = 0$ with $\underline{1} = [1, \ldots, 1]^T \in R^{N \times 1}$. In addition, for agent i, its neighbor set can be described as $N_i = \{j | (\mathscr{V}_j, \mathscr{V}_i) \in E\}$. Obviously, for the $i - th$ agent, if the $j - th$ agent is its neighbor, agent i can obtain the information from the $j - th$ agent. For directed graph, however, not necessarily vice versa. For undirected graph, neighbors are interrelated. From node i to node j, if there exists a sequence of successive arcs, which can be described in the form $\{(\mathscr{V}_i, \mathscr{V}_l), (\mathscr{V}_l, \mathscr{V}_k), \ldots, (\mathscr{V}_m, \mathscr{V}_j)\}$, then a direct path exists. In a digraph, if there exists a special node and a directed path from the special node to every other node, then graph is said to have a spanning tree. For any two different nodes $\mathscr{V}_i$ and $\mathscr{V}_j$ in the digraph, from $\mathscr{V}_i$ to $\mathscr{V}_j$, if there always exists a direct path, then the graph is said to be strongly connected.

Let $\bar{\mathscr{E}} \subseteq \bar{\mathscr{V}} \times \bar{\mathscr{V}}$ and $\bar{\mathscr{V}} = \{\mathscr{V}_0, \mathscr{V}_1, \mathscr{V}_2, \ldots, \mathscr{V}_N\}$. Then, based on the graph $\mathscr{G}$ and a leader node, a new graph $\bar{\mathscr{G}}$ is defined as $\bar{\mathscr{G}} = \{\bar{\mathscr{V}}, \bar{\mathscr{E}}\}$. Note that, from the leader node to node k, if there exists an edge, then its weight $b_k > 0$, otherwise $b_k = 0$, where $k = 1, 2, \ldots, N$.

Now, the following lemma and Assumption are made.

Assumption 7.1 There exists a spanning tree in graph $\bar{\mathscr{G}}$, where the leader is the root, and its topology is invariant.

Lemma 7.1 (Zhang and Lewis 2005) *Define* $q = (\mathscr{B} + \mathscr{L})^{-1}\underline{1} = [q_1, \ldots, q_N]$,

$$P = diag\{\frac{1}{q_i}\} \in R^{N \times N} \ \ with \ \ p_i = \frac{1}{q_i}, \quad Q = P(\mathscr{B} + \mathscr{L}) + (\mathscr{B} + \mathscr{L})^T P,$$

then matrices Q *and* P *are positive definite, i.e.,* $Q > 0$ *and* $P > 0$.

7.2.2 *Problem Formulation*

In this chapter, let us consider a leader-following systems, which contains one leader and N followers. For follower k ($k = 1, 2, \ldots, N$), it has the following form

$$\begin{cases} \dot{z}_{k,1}(t) = z_{k,2}(t) \\ \vdots \\ \dot{z}_{k,n_k-1} = z_{k,n_k} \\ \dot{z}_{k,n_k}(t) = f_k(\bar{z}_k) + u_k(t) + h_k(t) \end{cases} \tag{7.1}$$

where order number n_k is a positive integer, $\bar{z}_k = [z_{k,1}, \ldots, z_{k,n_k}]^T \in R^{n_k}$ denotes the state vector with $z_{k,i} \in R$ being the ith state; control input $u_k \in R$; function $f_k(\cdot) \in R$ is continuous and unknown; external disturbance $h_k(t) \in R$ is bounded but unknown. The leader is presented as

$$\begin{cases} \dot{z}_{0,1}(t) = z_{0,2}(t) \\ \vdots \\ \dot{z}_{0,n_0-1} = z_{0,n_0} \\ \dot{z}_{0,n_0}(t) = f_0(\bar{z}_0, t) \end{cases} \tag{7.2}$$

where order number n_0 is a positive integer, state vector $\bar{z}_0 = [z_{0,1}, \ldots, z_{0,n}]^T \in R^{n_0}$ with $z_{0,i} \in R$ being the ith state; function $f_0 \in R$ with $f_0(0, t) = 0$, which is unknown for the followers.

In this chapter, in order to decrease the complexity, let us assume $n_0 = n_k = n$, $k = 1, 2, \ldots, N$.

Define $z_i = [z_{1,i}, \ldots, z_{N,i}]^T$. Then, (7.1) can be written as follows:

$$\begin{cases} \dot{z}_1(t) = z_2(t) \\ \vdots \\ \dot{z}_{n-1}(t) = z_n(t) \\ \dot{z}_n(t) = f(\bar{z}) + u(t) + h(t) \end{cases}, \quad i = 1, \ldots, n-1 \tag{7.3}$$

where $f = [f_1, \ldots, f_N]^T = [f_1(\bar{x}_1), \ldots, f_N(\bar{x}_N)]^T$, $u = [u_1, \ldots, u_N]^T$, $\bar{x} = [\bar{z}_1^T, \ldots, \bar{z}_N^T]^T$, $h = [h_1, \ldots, h_N]^T = [h_1(t), \ldots, h_N(t)]^T$.

In fault-free case, for follower k ($k = 1, \ldots, N$), the ith tracking error is defined as follows:

$$\delta_{k,i} = z_{k,i} - z_{0,i}, \quad i = 1, \ldots, n. \tag{7.4}$$

Define $\delta_i = [\delta_{1,i}, \ldots, \delta_{N,i}]^T$, one further has

$$\delta_i = z_i - \underline{z}_{0,i} \tag{7.5}$$

where vector $\underline{z}_{0,i} = [z_{0,i}, \ldots, z_{0,i}]^T \in R^N$.

The main task in this chapter is to propose a distributed control strategy such that the synchronization between the followers and leader is achieved with all the tracking errors δ_i asymptotically converge to small neighborhoods of the origin.

Now, the following neighborhood synchronization error is defined for follower k ($k = 1, 2, \ldots, N$)

$$e_{k,i}(t) = \sum_{j \in N_k} a_{kj}(z_{j,i} - z_{k,i}) + b_k(z_{0,i} - z_{k,i}) \tag{7.6}$$

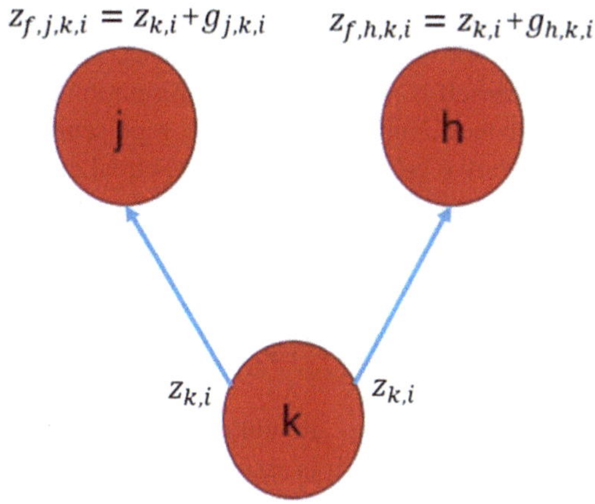

Fig. 7.1 Information transmission in the faulty case

Now, it is necessary to point out, $b_k > 0 \in R$ is a weight on the edge between the leader and follower k. As stated in the previous subsection, if follower k is able to access its information from the leader, namely, there exists an edge from the leader to follower k, then $b_k > 0$; otherwise, $b_k = 0$.

The following symbols are defined as follows:

$$B = diag\{b_1, \ldots, b_N\} \in R^{N \times N}.$$

$$e_i = [e_{1,i}, \ldots, e_{N,i}]^T \in R^N,$$

$$\underline{f}_0 = [f_0(x_0, t), \ldots, f_0(x_0, t)]^T \in R^N,$$

Tracking error dynamics (7.6) can be presented in the following compact form:

$$\begin{cases} \dot{e}_i(t) = e_{i+1}(t) \\ \vdots \\ \dot{e}_n(t) = -(B + L)(f + u(t) + h - \underline{f}_0) \end{cases} \tag{7.7}$$

Define $\bar{\upsilon} = \{\upsilon_0, \upsilon_1, \ldots, \upsilon_N\}$ and $\bar{\mathscr{E}} \subseteq \bar{\upsilon} \times \bar{\upsilon}$. Then, for the leader-following systems presented by (7.1) and (7.2), it can be modeled as a graph $\bar{\mathscr{G}} = \{\bar{\upsilon}, \bar{\mathscr{E}}\}$.

It is worth pointing out, in a communication network, if there is direct data communication or signal transmission between two nodes, there exists at least one communication link between them. In the following, Fig. 7.1is taken as an example to explain it, which has three nodes and two edges (communication links). One is from node k to node j, the other is from node k to node h. It means that node k is the neighbor of nodes j and j, and further implies that nodes j and j are able to access the leader's information via the corresponding communication links, respectively.

Without losing the generality, it is assumed that the information transmitted by the network is the states of the nodes, i.e., $z_{k,i}$. For node k, the information sent from himself is $z_{k,i}$. While, the information obtained by nodes j and h from node k through the communication links is represented by $z_{f,j,k,i}$ and $z_{f,h,k,i}$. In the fault-free case, the two information should be equal, namely, $z_{f,k,i} = z_{k,i}$. In real engineering, however, the links in the communication network maybe faulty. Just due to the communication fault, $z_{f,j,k,i}$ and $z_{f,h,k,i}$ are polluted, and $z_{f,j,k,i} \neq z_{k,i}$ and $z_{f,h,k,i} \neq z_{k,i}$. The polluted information can be described as follows:

$$z_{f,j,k,i} = z_{k,i} + g_{j,k,i}, \quad z_{f,h,k,i} = z_{k,i} + g_{h,k,i}. \tag{7.8}$$

where $g_{j,k,i}$ and $g_{h,k,i}$ are unknown bounded signal communication faults.

For convenience, it is assumed that $g_{j,k,i} = g_{h,k,i}$ in this chapter. Then, one has

$$z_{f,j,k,i} = z_{k,i} + g_{k,i}, \quad z_{f,h,k,i} = z_{k,i} + g_{k,i},$$

where $g_{k,i}$ is an unknown bounded faults. Obviously, $z_{f,j,k,i} = z_{f,h,k,i}$. Hence, as shown in Fig. 7.1, in the following, $z_{f,j,k,i}$ and $z_{f,h,k,i}$ are expressed as $z_{f,k,i}$.

Similarly, for the leader, the polluted information can be expressed as

$$z_{f,0,i} = z_{0,i} + g_{0,i}$$

where $g_{0,i}$ denotes an unknown bounded fault in communication link.

Let $g_k = [g_{k,1}, \dots, g_{k,n_k}]^T$, $z_f = [z_{f,1}^T, \dots, z_{f,N}^T]^T$, $g = [g_1^T, \dots, g_N^T]^T$, $z_{f,k} = [z_{f,k,1}, \dots, z_{f,k,n_k}]^T$. Then, one has

$$x_{f,k} = \bar{x}_k + g_k \quad z_f = \bar{z} + g.$$

In the faulty case, for the neighborhood synchronization error in (7.6), its actual value is expressed as $e_{a,k,i}$, which has the following form:

$$\begin{aligned} e_{a,k,i} &= \sum_{j \in N_i} a_{kj}(z_{f,j,i} - z_{f,k,i}) + b_k(z_{f,0,i} - z_{f,k,i}) \\ &= \sum_{j \in N_i} a_{kj}[(z_{j,i} + g_{j,i}) - (z_{k,i} + g_{k,i})] \\ &+ b_k[-(z_{k,i} + g_{k,i}) + (z_{0,i} + g_{0,i})] \\ &= e_{k,i} + b_k(g_{0,i} - g_{k,i}) + \sum_{j \in N_i} a_{kj}(g_{j,i} - g_{k,i}) \end{aligned} \tag{7.9}$$

Define

$$\omega_k = b_k(g_{0,1} - g_{k,1}) + \sum_{j \in N_1} a_{kj}(g_{j,1} - g_{k,1}),$$

then

$$e_{a,k,1} = \omega_k + e_{k,1} \tag{7.10}$$

Now, let us make some assumptions.

Assumption 7.2 $\bar{z}_0(t)$ and $f_0(\cdot, t)$ are bounded, i.e., $||\bar{z}_0(t)|| \leq M_0$, $|f_0(\cdot, t)| \leq M_{f0}$ with $M_0 \in R$ and $M_{f0} \in R$ being positive constants.

Assumption 7.3 $h_k(\cdot, t)$ is bounded, i.e., $|h_k(\cdot, t)| \leq M_{h,k}$ with $M_{h,k}$ being a positive constant.

Assumption 7.4 ω_k is bounded, i.e., $|\omega_k| \leq M_{\omega,k}$ with $M_{\omega,k}$ being a positive constant.

Lemma 7.2 (Moreau 2005) $||\varphi_i|| \leq \frac{1}{\underline{\varsigma}(B+L)}||e_{a,i}||$, *where* $\underline{\varsigma}(B+L)$ *is a minimum singular value of matrix* $B + L$, $i = 1, \ldots, n$,.

7.2.3 *Neural Networks*

As stated in Chap. 2, as one of the universal approximations, neural networks (NNs) have been utilized widely in control design of dynamic systems, especially uncertain nonlinear systems. Radial basis function (RBF) NNs shown in Fig. 7.2, as special NNs, will be approximated by the unknown continuous function Shen et al. (2014), Jiao et al. (2018; 2018). The detailed description of RBF NNs can be found in Chap. 2 and omitted here.

In this chapter, NNs will be used to approximate the unknown functions in (7.1), i.e., $f_k(z_k))$. Over the region $\Omega_{z,k}$, by using the above neural networks, it is represented as

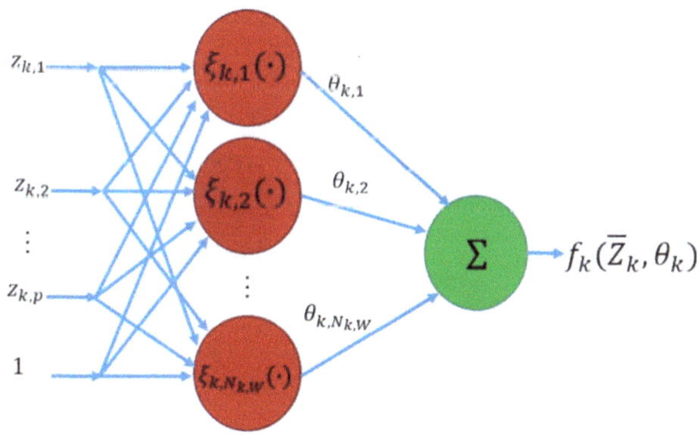

Fig. 7.2 RBF neural networks (in the fault-free case)

$$f_k(z_k) = \theta_k^T \zeta_k(\bar{Z}_k) + \varepsilon_k(\bar{Z}_k), \tag{7.11}$$

where $\bar{Z}_k = [z_k^T, 1]^T = [Z_{k,1}, \ldots, Z_{k,p_k}]^T$, approximation error $\varepsilon_k(\bar{Z}_k) \in R$, $\zeta_k(\bar{Z}_k) = \left[\zeta_{k,1}(\bar{Z}_k), \ldots, \zeta_{k,N_{k,W}}(\bar{Z}_k)\right]^T$ with $N_{k,W}$ being the number of nodes in the NNs, $\zeta_{k,i}(\bar{Z}_k) = \exp(-\sum_{j=1}^{\psi_k} \frac{(Z_{k,j}-q_{k,i,j})^2}{c_{k,i}{}^2})$, ψ_k denotes the dimension of $\bar{Z}_k$, $q_{k,i,j}$ is the width of the Gaussian function and $c_{k,i} > 0$ is the center of the receptive field.

Now, let us define the following set, optimal weight vector, and approximation error as follows:

$$\Omega_\theta = \{\theta_k \,||\, \|\theta_k\| \leq \beta_\theta\}$$

$$\theta_k^* = \arg\min_{\theta \in \Omega_\theta}[\sup_{\bar{z} \in \Omega_z} \left|\theta_k^T \zeta_k(\bar{Z}_k) - f_k(\bar{z}_k)\right|]$$

$$\varepsilon_k(\bar{Z}_k) = f_k(\bar{z}_k) - \theta_k^{*T} \zeta_k(\bar{Z}_k)$$

with $\beta_\theta > 0$ is a real constant, Ω_z is a compact set, which can be enough large.

From NNs' function approximation theory Shen et al. (2014), one has, on a compact set, any continuous function that can be approximated by neural networks as precisely as possible. Thus, the optimal approximation error $\varepsilon_k(\bar{Z}_k)$ is bounded although it is unknown.

From Fig. 7.2, one can find that for an NN, its input vector contains system state vector. In this chapter, however, since signal communication faults are considered, the actual value of the state z_k cannot be obtained, only the polluted value, i.e., $z_{f,k}$, is accessible. Thus, when using the neural network to approximate $f_k(\bar{z}_k)$ in (7.1), the communication faults should be considered. From Fig. 7.3, one can find, it is $z_{f,k}$ not z_k that is used in neural networks. From the above analysis, one has, a function $f_k(\bar{z}_k)$ can be approximated as

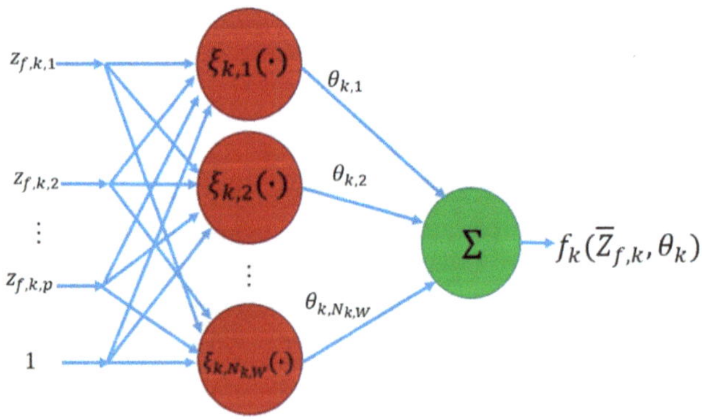

Fig. 7.3 RBF neural networks (in the faulty case)

$$f_k(\bar{z}_k) = \theta_k^T \zeta_k(\bar{Z}_k) + \varepsilon_k(\bar{Z}_k).$$

Now, the term $\theta_k^{*T} \zeta_k(\bar{Z}_{f,k})$ is added to and subtracted from the right hand of the above equation. Further, one has

$$\begin{aligned} f_k(\bar{z}_k) &= \theta_k^{*T} \zeta_k(\bar{Z}_k) + \varepsilon_k + \theta_k^{*T} \zeta_k(\bar{Z}_{f,k}) - \theta_k^{*T} \zeta_k(\bar{Z}_{f,k}) \\ &= \theta_k^{*T} \zeta_k(\bar{Z}_{f,k}) + \varepsilon_k - \theta_k^{*T} [\zeta_k(\bar{Z}_{f,k}) - \zeta_k(\bar{Z}_k)] \end{aligned} \tag{7.12}$$

Assumption 7.5 Approximation error ε_k and optimal weight vector θ_k are bounded, namely, $|\varepsilon_k| \le M_{\varepsilon,k}$ and $||\theta_k|| \le \beta_\theta$ with $M_{\varepsilon,k}$ and β_θ being positive real constants.

7.3 Main Results

For follower k, sliding mode error is defined as

$$s_k = (\frac{d}{dt} + \lambda)^{n-1} e_{k,1}(t) = \sum_{i=1}^{n-1} c_{k,i} e_{k,i}(t) + e_{k,n}(t), k = 1, \ldots, N \tag{7.13}$$

where $c_{k,i} = C_{n-1}^{i-1} \lambda_k^{n-i}$ with $\lambda_k > 0$ being a design parameter.

Define $\bar{e}_k(t) = [e_{k,1}(t), \ldots, e_{k,n}(t)]^T$. Then, a valuable lemma is introduced.

Lemma 7.3 (Shen et al. 2019) *Let s_k be defined by (7.13). Then one has*
(1) if $s_k = 0$, then $\lim_{t\to\infty} \bar{e}_k(t) = 0$;
(2) if $|s_k| \le \alpha_k$ and $\bar{e}_k(0) \in \Omega_{\alpha_k}$, then $e_k(t) \in \Omega_{\alpha_k}, \forall t \ge 0$;
(3) if $|s_k| \le \alpha_k$ and $\bar{e}_k(0) \notin \Omega_{\alpha_k}$, then $\exists T_k = (m_k - 1)/\lambda_k$, for $\forall t \ge T_k$, $\bar{e}_k(t) \in \Omega_{\alpha_k}$,
where $\Omega_{\alpha_k} = \{\bar{e}_k(t) |\, |e_{k,i}| \le 2^{(j-1)} \lambda_k^{j-m_k} \alpha_k,\ j = 1, \ldots m_k\}$, $i = 1, \ldots, n$.

In order to decrease the complexity, it is set that $c_{l,i} = \lambda_i$ and $\lambda_n = 1$, $l = 1, 2, \ldots, N$, then

$$s_k = \lambda_1 e_{k,1} + \cdots + \lambda_n e_{k,n} \tag{7.14}$$

Define the global sliding mode error $s = [s_1, \ldots, s_N]^T$, then

$$s = \lambda_1 e_1 + \cdots + \lambda_n e_n \tag{7.15}$$

Differentiating s, we have

$$\begin{aligned} \dot{s} &= \lambda_1 \dot{e}_1 + \cdots + \lambda_n \dot{e}_n = \sum\nolimits_{i=1}^{n-1} \lambda_i e_i + \dot{e}_n \\ &= \gamma - (\mathscr{B} + \mathscr{L})(f(\bar{x}) + u + h - \underline{f}_0) \end{aligned} \tag{7.16}$$

where $\gamma = \sum_{i=1}^{n-1} \lambda_i e_{i+1}$.

Let $\zeta_f = [\zeta_{f,1}, \ldots, \zeta_{f,N}]^T$ with $\zeta_{f,k} = \theta_k^{*T}[\zeta(\bar{Z}_{f,k}) - \zeta(\bar{Z}_k)]$, $\theta^{*T} = diag\{\theta_1^{*T}, \ldots, \theta_N^{*T}\}$, $\zeta = [\zeta_1^T(\bar{Z}_{f,1}), \ldots, \zeta_N^T(\bar{Z}_{f,N})]^T$, $\varepsilon = [\varepsilon_1, \ldots, \varepsilon_N]^T$, then

$$f(\bar{x}) = \theta^{*T}\zeta + \varepsilon - \zeta_f$$

If the communication faults occur in the data transmission links, s_k is polluted, which is expressed as $s_{f,k}$. As stated in the previous subsection, only the polluted signal $s_{f,k}$ can be used in control design. Similar to (7.14), $s_{f,k}$ can be described as

$$\begin{aligned} s_{f,k} &= (\frac{d}{dt} + \lambda)^{n-1} e_{a,k,1}(t) \\ &= \sum_{i=1}^{n-1} c_{k,i} e_{k,i}(t) + e_{k,n}(t) \\ &\quad + \lambda_1[\sum_{j \in N_i} a_{kj}(g_{f,j,1} - g_{f,k,1}) + b_k(g_{f,0,1} - g_{f,k,1})] \\ &= s_k + \lambda_1 \omega_k \end{aligned} \tag{7.17}$$

Further, one has

$$s_f = s + \omega \tag{7.18}$$

where $s_f = [s_{f,1}, s_{f,2}, \ldots, s_{f,n}]^T$, $\omega = [\lambda_1\omega_1, \ldots, \lambda_1\omega_n]^T$.

Differentiating s_f with respect to time t, we have

$$\dot{s}_f = \dot{s} + \dot{\omega} = \gamma - (\mathscr{B} + \mathscr{L})(f(\bar{x}) + u + h - \underline{f}_0) \tag{7.19}$$

Define

$$V_s = s^T P s/2 \tag{7.20}$$

here $P = P^T > 0 \in R^{n\times n}$ is defined in Lemma 7.1.

Differentiating V_s with respect to time t, we have

$$\begin{aligned} \dot{V}_s &= s^T P\left[\gamma - (\mathscr{B} + \mathscr{L})(\theta^{*T}\zeta - \zeta_f + \varepsilon + u - \underline{f}_0 + h)\right] \\ &= s^T P\gamma - s^T P(\mathscr{B} + \mathscr{L})u - s^T P(\mathscr{B} + \mathscr{D})\theta^{*T}\zeta \\ &\quad + s^T PA\theta^{*T}\zeta + s^T P(\mathscr{B} + \mathscr{D})\zeta_f - s^T PA\zeta_f \\ &\quad - s^T P(\mathscr{B} + \mathscr{D})(\varepsilon - \underline{f}_0 + h) + s^T PA(\varepsilon - \underline{f}_0 + h) \end{aligned} \tag{7.21}$$

Control law is defined as follows:

$$u = (\mathscr{B} + \mathscr{D})^{-1}\gamma - \hat{f} + m(s_f - \hat{\omega}) - \mathrm{sgn}((s_f - \hat{\omega})^T P(\mathscr{B} + \mathscr{D}))\hat{\bar{M}}_{\varepsilon hf} \tag{7.22}$$

where $\hat{f}_k = \hat{\theta}_k^T \zeta_k(\bar{Z}_{f,k})$ and $\hat{M}_{\varepsilon hf,k}$ denote respectively the estimates of $f_k(\bar{Z}_{f,k})$ and $M_{\varepsilon hf}$, $M_{\varepsilon hf} = M_{\varepsilon,k} + M_{h,k} + M_{f0}$, $\hat{f} = [\hat{f}_1, \ldots, \hat{f}_N]^T$, $\hat{\bar{M}}_{\varepsilon hf} = [\hat{\bar{M}}_{\varepsilon hf,1}, \ldots, \hat{\bar{M}}_{\varepsilon hf,N}]^T$, design parameter $m > 0 \in R$ is chosen such that

$$-r\bar{\varsigma}(P) - (5r + 2rm + \frac{1}{4r\underline{\lambda}^2}\bar{\lambda}^2)\varsigma(A)\bar{\varsigma}(P) + \frac{1}{2}m\underline{\varsigma}(Q) > 0$$

Define the notations as follows:

$$\tilde{\theta} = \theta^* - \hat{\theta}, \quad \theta^T = diag(\theta_1^T, \ldots, \theta_N^T), \quad \hat{\theta}^T = diag(\hat{\theta}_1^T, \ldots, \hat{\theta}_N^T),$$
$$\tilde{\bar{M}}_{\varepsilon hf} = \bar{M}_{\varepsilon hf} - \hat{\bar{M}}_{\varepsilon hf}, \quad \tilde{\omega} = \omega - \hat{\omega}$$

with $\hat{\theta}_k^T$ being the estimates of θ_k^{*T}.

Substituting the control law (7.22) into (7.21), one has

$$\begin{aligned}
\dot{V}_s =& -ms^T P(\mathscr{B} + \mathscr{D})s + ms^T PAs - s^T P(\mathscr{B} + \mathscr{D})\tilde{\theta}^T \zeta \\
& - s^T P(\mathscr{B} + \mathscr{D})\text{sgn}(s^T P(\mathscr{B} + \mathscr{D}))\tilde{\bar{M}}_{\varepsilon hf} + s^T PA\bar{M}_{\varepsilon hf} \\
& + ms^T PA\omega - ms^T PA\hat{\omega} + s^T PA\tilde{\theta}^T \zeta - s^T PA\zeta_f \\
& - ms^T P(\mathscr{B} + \mathscr{D})\tilde{\omega} - s^T PA\text{sgn}(s^T P(\mathscr{B} + \mathscr{D}))\hat{\bar{M}}_{\varepsilon hf} \\
& + s^T PA(\mathscr{B} + \mathscr{D})^{-1}\gamma + s^T P(\mathscr{B} + \mathscr{D})\zeta_f
\end{aligned} \tag{7.23}$$

Since

$$-s^T PA\zeta_f \leq \bar{\varsigma}(P)\bar{\varsigma}(A)rs^T s + \frac{1}{4r}\bar{\varsigma}(A)\bar{\varsigma}(P)\zeta_f^T \zeta_f$$

$$s^T PA\tilde{\theta}^T \zeta \leq \bar{\varsigma}(A)\bar{\varsigma}(P)rs^T s + \frac{1}{4r}\bar{\varsigma}(A)\bar{\varsigma}(P)\zeta^T \tilde{\theta}\tilde{\theta}^T \zeta$$

$$ms^T PA\omega \leq \bar{\varsigma}(A)\bar{\varsigma}(P)rms^T s + \frac{1}{4r}\bar{\varsigma}(A)\varsigma(P)m\omega^T \omega$$

$$-ms^T PA\hat{\omega} \leq \bar{\varsigma}(A)\bar{\varsigma}(P)rms^T s + \frac{1}{4r}\bar{\varsigma}(A)\bar{\varsigma}(P)m\beta_\omega^2$$

$$s^T PA\bar{M}_{\varepsilon hf} \leq \bar{\varsigma}(A)\bar{\varsigma}(P)rs^T s + \frac{1}{4r}\bar{\varsigma}(A)\bar{\varsigma}(P)\bar{M}_{\varepsilon hf}^T \bar{M}_{\varepsilon hf}$$

$$s^T PA(\mathscr{B} + \mathscr{D})^{-1}\gamma \leq \bar{\varsigma}(A)\bar{\varsigma}(P)rs^T s + \frac{1}{4r}\bar{\varsigma}(A)\bar{\varsigma}(P)\gamma^T \gamma$$

$$s^T P(\mathscr{B} + \mathscr{D})\zeta_f \leq \bar{\varsigma}(P)rs^T s + \frac{1}{4r}\bar{\varsigma}(P)\zeta_f^T (\mathscr{B} + \mathscr{D})^T (\mathscr{B} + \mathscr{D})\zeta_f$$

$$- s^T P A \text{sgn}(s^T P(\mathscr{B}+\mathscr{D}))\hat{\bar{M}}_{\varepsilon hf} \leq \bar{\varsigma}(A)\bar{\varsigma}(P) r s^T s + \frac{1}{4r}\bar{\varsigma}(A)\bar{\varsigma}(P)\beta_M^2$$

where design parameter $r > 0 \in R$, and $\bar{M}_{\varepsilon hf} = [M_{\varepsilon hf,1}, \ldots, M_{\varepsilon hf,N}]^T$, it follows

$$\begin{aligned}\dot{V}_s \leq & - m s^T P(\mathscr{B}+\mathscr{L})s - s^T P(\mathscr{B}+\mathscr{D})\tilde{\theta}^T \xi - m s^T P(\mathscr{B}+\mathscr{D})\tilde{\omega} \\ & - s^T P(\mathscr{B}+\mathscr{D})\text{sgn}(s^T P(\mathscr{B}+\mathscr{D}))\tilde{\bar{M}}_{\varepsilon hf} + \bar{\varsigma}(A)\bar{\varsigma}(P)/4r\gamma^T\gamma \\ & + r\bar{\varsigma}(P)(5\bar{\varsigma}(A) + 2m\bar{\varsigma}(A) + 1)s^T s + \bar{\varsigma}(P)/4r\zeta_f^T(\mathscr{B}+\mathscr{D})^T(\mathscr{B}+\mathscr{D})\zeta_f \\ & + \bar{\varsigma}(A)\bar{\varsigma}(P)/4r(\zeta^T\tilde{\theta}\tilde{\theta}^T\zeta + \zeta_f^T\zeta_f + \beta_M^2 + m\beta_\omega^2 + m\omega^T\omega + \bar{M}_{\varepsilon hf}^T\bar{M}_{\varepsilon hf})\end{aligned} \tag{7.24}$$

From Assumption 7.5, it follows that θ_k^* is bounded. The adaptive law (7.30) guarantees that $\hat{\theta}_k$ is also bounded. Therefore, one obtains that $\tilde{\theta}^T\zeta$ is bounded. In addition, from Assumptions 7.2, 7.3 and 7.5, it follows that $\bar{M}_{h\varepsilon f}$ is bounded. The adaptive law (7.31) also guarantees that $\hat{\bar{M}}_{h\varepsilon f}$ is bounded. Thus, one can obtain $\tilde{\bar{M}}_{h\varepsilon f}$ is bounded.

Due to the boundedness of $\tilde{\theta}^T\xi$, $\bar{M}_{h\varepsilon f}$ and ω, there must be a parameter r that makes the following conditions true,

$$\frac{1}{4r}\bar{\varsigma}(A)\bar{\varsigma}(P)(\zeta^T\tilde{\theta}\tilde{\theta}^T\zeta + \zeta_f^T\zeta_f + \beta_M^2 + m\beta_\omega^2 + m\omega^T\omega + \bar{M}_{\varepsilon hf}^T\bar{M}_{\varepsilon hf}) \leq \mu_{s1}$$

$$\frac{1}{4r}\bar{\varsigma}(P)\zeta_f^T(\mathscr{B}+\mathscr{D})^T(\mathscr{B}+\mathscr{D})\zeta_f \leq \mu_{s2}.$$

Since

$$\gamma_k^2 = \sum_{i=1}^{n-1}\lambda_i^2 e_{k,i+1}^2 \leq \bar{\lambda}^2 s_k^2/\underline{\lambda}^2$$

$$\gamma^T\gamma = \sum_{k=1}^{N}\gamma_k^2 \leq \sum_{k=1}^{N}\frac{\bar{\lambda}^2}{\underline{\lambda}^2}s_k^2 = \frac{\bar{\lambda}^2}{\underline{\lambda}^2}\sum_{k=1}^{N}s_k^2 = \frac{\bar{\lambda}^2}{\underline{\lambda}^2}s^T s$$

where $\bar{\lambda} = \max\{\lambda_1, \ldots, \lambda_n\}$, $\underline{\lambda} = \min\{\lambda_1, \ldots, \lambda_n\}$, (7.23) is developed as follows:

$$\begin{aligned}\dot{V}_s \leq & - m s^T P(\mathscr{B}+\mathscr{L})s + (\frac{1}{4r\underline{\lambda}^2}(5r + 2rm + \bar{\lambda}^2)\bar{\varsigma}(A) + r)\bar{\varsigma}(P)s^T s \\ & - s^T P(\mathscr{B}+\mathscr{D})\tilde{\theta}^T\zeta - s^T P(\mathscr{B}+\mathscr{D})\text{sgn}(s^T P(\mathscr{B}+\mathscr{D}))\tilde{\bar{M}}_{\varepsilon hf} \\ & - m s^T P(\mathscr{B}+\mathscr{D})\tilde{\omega} + \mu_s\end{aligned} \tag{7.25}$$

where $\mu_s = \mu_{s1} + \mu_{s2}$.

Define

$$V_\theta = \tilde{\theta}^T\tilde{\theta}/2\eta_1 + \tilde{\bar{M}}_{\varepsilon hf}^T\tilde{\bar{M}}_{\varepsilon hf}/2\eta_2 + \tilde{\omega}^T\tilde{\omega}/2\eta_3 \tag{7.26}$$

where design parameters $\eta_1 > 0 \in R$, $\eta_2 > 0 \in R$, $\eta_3 > 0 \in R$.

Differentiating V_θ with respect to time t , we have

$$\dot{V}_\theta = -\tilde{\theta}^T \dot{\hat{\theta}}/\eta_1 - \tilde{\bar{M}}^T_{\varepsilon hf} \dot{\hat{\bar{M}}}_{\varepsilon hf}/\eta_2 - \tilde{\omega}^T \dot{\hat{\omega}}/\eta_3 \tag{7.27}$$

Define

$$V_0 = V_s + V_\theta \tag{7.28}$$

For V_0, its time derivative is

$$\begin{aligned}\dot{V}_0 \le & - m s^T P(\mathscr{B}+\mathscr{L})s + (\frac{1}{4r\underline{\lambda}^2}(5r + 2rm + \bar{\lambda}^2)\bar{\varsigma}(A) + r)\bar{\varsigma}(P)s^T s \\ & - s^T P(\mathscr{B}+\mathscr{D})\tilde{\theta}^T \zeta - s^T P(D+B)\mathrm{sgn}(s^T P(\mathscr{B}+\mathscr{D}))\tilde{\bar{M}}_{\varepsilon hf} \\ & - m s^T P(\mathscr{B}+\mathscr{D})\tilde{\omega} - \tilde{\theta}^T \dot{\hat{\theta}}/\eta_1 - \tilde{\bar{M}}^T_{\varepsilon hf} \dot{\hat{\bar{M}}}_{\varepsilon hf}/\eta_2 - \tilde{\omega}^T \dot{\hat{\omega}}/\eta_3 + \mu_s\end{aligned} \tag{7.29}$$

Adaptive laws are defined as follows:

$$\dot{\hat{\theta}} = \begin{cases} \tau_\theta, & \text{if } ||\hat{\theta}|| < \beta_\theta \text{ or } ||\hat{\theta}|| = \beta_\theta \text{ and } \hat{\theta}^T \tau_\theta \le 0 \\ \tau_\theta - \dfrac{\hat{\theta}\hat{\theta}^T}{||\hat{\theta}||^2}\tau_\theta, & \text{if } ||\hat{\theta}|| = \beta_\theta \text{ and } \hat{\theta}^T \tau_\theta > 0 \end{cases} \tag{7.30}$$

$$\dot{\hat{\bar{M}}}_{\varepsilon hf} = \begin{cases} \tau_M, if||\hat{\bar{M}}_{\varepsilon hf}|| < \beta_M or ||\hat{\bar{M}}_{\varepsilon hf}|| = \beta_M \text{ and } \hat{\bar{M}}^T_{\varepsilon hf}\tau_M \le 0 \\ \tau_M - \dfrac{\hat{\bar{M}}_{\varepsilon hf}\hat{\bar{M}}^T_{\varepsilon hf}}{||\hat{\bar{M}}_{\varepsilon hf}||^2}\tau_M, if||\hat{\bar{M}}_{\varepsilon hf}|| = \beta_M \text{ and } \hat{\bar{M}}^T_{\varepsilon hf}\tau_M > 0 \end{cases} \tag{7.31}$$

$$\dot{\hat{\omega}} = \begin{cases} \tau_\omega, & \text{if } ||\hat{\omega}|| < \beta_\omega \text{ or } ||\hat{\omega}|| = \beta_\omega \text{ and } \hat{\omega}^T \tau_\omega \le 0 \\ \tau_\omega - \dfrac{\hat{\omega}\hat{\omega}^T}{||\hat{\omega}||^2}\tau_\omega, & \text{if } ||\hat{\omega}|| = \beta_\omega \text{ and } \hat{\omega}^T \tau_\omega > 0 \end{cases} \tag{7.32}$$

where $\tau_\theta = -\eta_1\zeta(s_f - \hat{\omega})^T P(\mathscr{B}+\mathscr{D}) + \eta_\theta\hat{\theta}$, $\tau_M = -\eta_2\mathrm{sgn}((s_f - \hat{\omega})^T P(\mathscr{B}+\mathscr{D}))(s_f - \hat{\omega})^T P(\mathscr{B}+\mathscr{D}) + \eta_M \hat{\bar{M}}_{\varepsilon hf}$, $\tau_\omega = -\eta_3 m\lambda_1(s_f - \hat{\omega})^T P(\mathscr{B}+\mathscr{D}) + \eta_\omega\hat{\omega}$. design parameters $\eta_\kappa > 0 \in R$, $\kappa = \theta, M, \omega$. Notice that, for $\hat{\theta}$, $\hat{\bar{M}}_{\varepsilon hf}$ and $\hat{\omega}$, the project operators in (7.30)–(7.32) guarantee their boundedness.

Now, the main results are given by the following theorem.

Theorem 7.1 *Under Assumptions 7.1–7.5, consider a leader-following system described by (7.1) and (7.2). If the distributed control law (7.22) and the adaptive laws (7.30)–(7.32) are applied, then the synchronization between the leader* *and followers is achieved with the tracking errors asymptotically converge to small neighborhoods of the origin.*

Proof Define

$$V_0 = \frac{1}{2}s^T P s + \frac{1}{2\eta_1}\tilde{\theta}^T\tilde{\theta} + \frac{1}{2\eta_2}\tilde{\tilde{M}}_{\varepsilon hf}^T\tilde{\tilde{M}}_{\varepsilon hf} + \frac{1}{2\eta_3}\tilde{\omega}^T\tilde{\omega}$$

Differentiating V_0 with respect to time t, we have

$$\begin{aligned}\dot{V}_0 =& s^T P\gamma - s^T P(\mathscr{B}+\mathscr{L})u - s^T P(\mathscr{B}+\mathscr{D})\theta^{*T}\zeta \\ &+ s^T PA\theta^{*T}\zeta + s^T P(\mathscr{B}+\mathscr{D})\zeta_f - s^T PA\zeta_f \\ &- s^T P(\mathscr{B}+\mathscr{D})(-\underline{f}_0 + h + \varepsilon) + s^T PA(-\underline{f}_0 + h + \varepsilon) \\ &- \tilde{\theta}^T\dot{\hat{\theta}}/\eta_1 - \tilde{\tilde{M}}_{\varepsilon hf}^T\dot{\hat{\tilde{M}}}_{\varepsilon hf}/\eta_2 - \tilde{\omega}^T\dot{\hat{\omega}}/\eta_3\end{aligned} \tag{7.33}$$

Substituting (7.22) and (7.30)–(7.32) into (33), it yields

$$\begin{aligned}\dot{V}_0 \leq& - ms^T P(\mathscr{B}+\mathscr{L})s + (\frac{1}{4r\underline{\lambda}^2}(5r + 2rm + \bar{\lambda}^2)\bar{\varsigma}(A) + r)\bar{\varsigma}(P)s^T s \\ &- \frac{\eta_\theta}{\eta_1}\tilde{\theta}^T\hat{\theta} - \frac{\eta_M}{\eta_2}\tilde{\tilde{M}}_{\varepsilon hf}^T\hat{\tilde{M}}_{\varepsilon hf} - \frac{\eta_\omega}{\eta_3}\tilde{\omega}^T\hat{\omega} + \mu_s\end{aligned} \tag{7.34}$$

Since

$$\begin{aligned}-\frac{\eta_\theta}{\eta_1}\tilde{\theta}^T\hat{\theta} &\leq -\frac{\eta_\theta}{2\eta_1}\tilde{\theta}^T\tilde{\theta} + \frac{\eta_\theta}{2\eta_1}\theta^{*T}\theta^* \\ -\frac{\eta_M}{\eta_2}\tilde{\tilde{M}}_{\varepsilon hf}^T\hat{\tilde{M}}_{\varepsilon hf} &\leq -\frac{\eta_M}{2\eta_2}\tilde{\tilde{M}}_{\varepsilon hf}^T\tilde{\tilde{M}}_{\varepsilon hf} + \frac{\eta_M}{2\eta_2}\bar{M}_{\varepsilon hf}^T\bar{M}_{\varepsilon hf} \\ -\frac{\eta_\omega}{\eta_3}\tilde{\omega}^T\hat{\omega} &\leq -\frac{\eta_\omega}{2\eta_3}\tilde{\omega}^T\tilde{\omega} + \frac{\eta_\omega}{2\eta_3}\omega^T\omega\end{aligned}$$

one has

$$\begin{aligned}\dot{V}_0 \leq& -\frac{1}{2}ms^T Qs + (\frac{1}{4r\underline{\lambda}^2}(5r + 2rm + \bar{\lambda}^2)\bar{\varsigma}(A) + r)\bar{\varsigma}(P)s^T s \\ &- \frac{\eta_\theta}{2\eta_1}\tilde{\theta}^T\tilde{\theta} - \frac{\eta_M}{2\eta_2}\tilde{\tilde{M}}_{\varepsilon hf}^T\tilde{\tilde{M}}_{\varepsilon hf} - \frac{\eta_\omega}{2\eta_3}\tilde{\omega}^T\tilde{\omega} \\ &+ \frac{\eta_\theta}{2\eta_1}\theta^{*T}\theta^* + \frac{\eta_M}{2\eta_2}\bar{M}_{\varepsilon hf}^T\bar{M}_{\varepsilon hf} + \frac{\eta_\omega}{2\eta_3}\omega^T\omega + \mu_s\end{aligned} \tag{7.35}$$

Since θ^*, $\bar{M}_{\varepsilon hf}$, ω are bounded, if η_θ, η_M, η_ω and η_1, η_2, η_3 can be appropriately chosen, there exists a design parameter $\mu_{\theta M} > 0 \in R$ such that

$$\frac{\eta_\theta}{2\eta_1}\theta^{*T}\theta^* + \frac{\eta_M}{2\eta_2}\bar{M}_{\varepsilon hf}^T\bar{M}_{\varepsilon hf} + \frac{\eta_\omega}{2\eta_3}\omega^T\omega \leq \mu_{\theta M}.$$

Let

$$\mu_0 = \mu_s + \mu_{\theta M},$$

$$\lambda_0 = \min\left\{\frac{1/2m\underline{\varsigma}(Q)}{\bar{\varsigma}(P)} - (5r + 2rm + \bar{\lambda}^2/4r\underline{\lambda}^2)\bar{\varsigma}(A) - r, \frac{\eta_\theta}{2\eta_1}, \frac{\eta_M}{2\eta_2}, \frac{\eta_\omega}{2\eta_3}\right\}.$$

then one has

$$\dot{V}_0 \leq -\lambda_0 V_0 + \mu_0.$$

Further, one has

$$0 \leq V_0(t) \leq \frac{\mu_0}{\lambda_0} + (V_0(t_0) - \frac{\mu_0}{\lambda_0})e^{-\lambda_0(t-t_0)}.$$

Let $V_0 = \frac{1}{2}\underline{\varsigma}(P)M^2$ with design parameter M being a positive real constant. Since

$$\underline{\varsigma}(P)||s(t)||^2 \leq 2V_s(t) \leq 2V_0(t),$$

then one has

$$||s(t)|| \leq \sqrt{2V_0(t)/\underline{\varsigma}(P)}, ||s(t)|| \leq M.$$

Similarly, one also has

$$||\tilde{\theta}(t)|| \leq \sqrt{\eta_1}M, ||\tilde{\bar{M}}_{\varepsilon hf}(t)|| \leq \sqrt{\eta_2}M, ||\tilde{\omega}|| \leq \sqrt{\eta_3}M.$$

From the above analysis, one has that s and ω are bounded. And because $s_f = s + \omega$, so s_f also is bounded.

Since $|e_{k,i}| \leq 2^{i-1}\lambda_k^{i-n}M$, one has

$$||e_i|| \leq \sqrt{N(2^{i-1}\lambda_k^{i-n}M)^2}.$$

From (7.16), one has, for $i = 1, \ldots, n$ and $k = 1, \ldots, N$,

$$||e_{a,i}|| \leq \sqrt{N(2^{i-1}\lambda_k^{i-n}M)^2} + \beta_\omega,$$

which means that the followers can asymptotically converge with the leader and the tracking and synchronization errors asymptotically converge, respectively, to the neighborhoods of the origin. The proof is completed here.

7.4 Simulation Results

Consider a leader-following system described in Fig. 7.4, which has one leader and five followers. The leader's dynamics is presented in the following form:

$$\begin{cases} \dot{z}_{0,1}(t) = z_{0,2}(t) \\ \dot{z}_{0,2}(t) = z_{0,3}(t) \\ \dot{z}_{0,3}(t) = z_{0,2}(t) - 2z_{0,3}(t) + 2\sin(2t) + 4\cos(2t) \end{cases}$$

and the followers have the same form of (7.1), namely,

$$\begin{cases} \dot{z}_{1,1}(t) = z_{1,2} \\ \dot{z}_{1,2}(t) = z_{1,3} \\ \dot{z}_{1,3}(t) = -z_{1,1}z_{1,2} + \sin(z_{1,3}) + u_1 + h_1 \end{cases}$$
$$\begin{cases} \dot{z}_{2,1}(t) = z_{2,2} \\ \dot{z}_{2,2}(t) = z_{2,3} \\ \dot{z}_{2,3}(t) = z_{2,1}\cos(z_{2,2}) + 2z_{2,3} + u_2 + h_2 \end{cases}$$
$$\begin{cases} \dot{z}_{3,1}(t) = z_{3,2} \\ \dot{z}_{3,2}(t) = z_{3,3} \\ \dot{z}_{3,3}(t) = -z_{3,1} + \sin(z_{3,2}) + u_3 + h_3 \end{cases}$$
$$\begin{cases} \dot{z}_{4,1}(t) = z_{4,2} \\ \dot{z}_{4,2}(t) = z_{4,3} \\ \dot{z}_{4,3}(t) = (z_{4,1} + z_{4,2})^2 + 3z_{4,3} + \cos(2t) + u_4 + h_4 \end{cases}$$
$$\begin{cases} \dot{z}_{5,1}(t) = z_{5,2} \\ \dot{z}_{5,2}(t) = z_{5,3} \\ \dot{z}_{5,3}(t) = -2z_{5,1} + z_{5,2} + u_5 + h_5 \end{cases}$$

In this chapter, the disturbance $|h_k| \leq 1$, which is randomly taken over the interval $[-1, 1]$.

The initial states are set as follows: $z_0 = [0, 1, 1]^T$, $z_1 = [1, -1, 0]^T$, $z_2 = [0, -1, -2]^T$, $z_3 = [0, 1, 0]^T$, $z_4 = [1.5, 0, 0]^T$, $z_5 = [0, 1, -1]^T$.

From Fig. 7.4, one has

$$A = \begin{bmatrix} 0&0&2&0&0 \\ 1&0&0&0&0 \\ 0&4&0&1&0 \\ 3&0&0&0&0 \\ 0&0&0&0&0 \end{bmatrix} \quad B = \begin{bmatrix} 5&0&0&0&0 \\ 0&0&0&0&0 \\ 0&0&0&0&0 \\ 0&0&0&0&0 \\ 0&0&0&0&5 \end{bmatrix} \quad D = \begin{bmatrix} 2&0&0&0&0 \\ 0&1&0&0&0 \\ 0&0&5&0&0 \\ 0&0&0&3&0 \\ 0&0&0&0&0 \end{bmatrix}$$

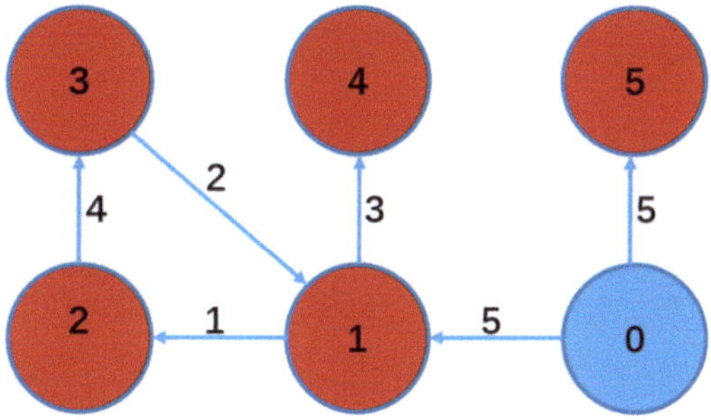

Fig. 7.4 Topology of the communication

$$L = \begin{bmatrix} 2 & 0 & -2 & 0 & 0 \\ -1 & 1 & 0 & 0 & 0 \\ 0 & -4 & 5 & -1 & 0 \\ -3 & 0 & 0 & 3 & 0 \\ 0 & 0 & 0 & 0 & 0 \end{bmatrix} \mathscr{B} + \mathscr{L} = \begin{bmatrix} 7 & 0 & -2 & 0 & 0 \\ -1 & 1 & 0 & 0 & 0 \\ 0 & -4 & 5 & -1 & 0 \\ -3 & 0 & 0 & 3 & 0 \\ 0 & 0 & 0 & 0 & 1 \end{bmatrix}$$

$$q = \begin{bmatrix} 0.62 \\ 1.62 \\ 1.69 \\ 0.96 \\ 1.00 \end{bmatrix} P = \begin{bmatrix} 1.59 & 0 & 0 & 0 & 0 \\ 0 & 0.61 & 0 & 0 & 0 \\ 0 & 0 & 0.59 & 0 & 0 \\ 0 & 0 & 0 & 1.04 & 0 \\ 0 & 0 & 0 & 0 & 1.00 \end{bmatrix}$$

$$Q = \begin{bmatrix} 11.16 & 0 & -1.18 & 0 & 0 \\ -1.59 & 0.61 & 0 & 0 & 0 \\ 0 & -2.45 & 2.95 & -1.04 & 0 \\ -4.78 & 0 & 0 & 3.12 & 0 \\ 0 & 0 & 0 & 0 & 1.00 \end{bmatrix}$$

Further, one has

$$\bar{\varsigma}(A) = 2.41, \bar{\varsigma}(P) = 1.59, \bar{\varsigma}(Q) = 11.01, \bar{\varsigma}(\mathscr{B} + \mathscr{D}) = 1.00.$$

In this simulation, design parameters are chosen as follows:

$$m = 1, \lambda_1 = 2, \lambda_2 = 2, \lambda_3 = 1, \eta_1 = \eta_2 = \eta_3 = 100, \eta_\theta = \eta_M = \eta_\omega = 0.01,$$

$$\mu_s = \mu_{\theta M} = 0.1, r = 100.$$

The simulation results are presented in Figs. 7.5 and 7.6. Figure 7.5 shows the boundedness of the leader states. From Fig. 7.6, one has, under the fault-free case, applying the control strategy proposed in this work, the synchronization between the leader and followers is quickly achieved, and the tracking errors converge asymptoti-

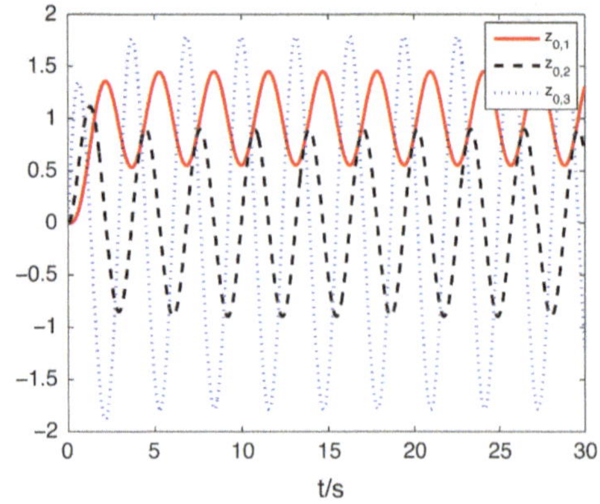

Fig. 7.5 The state of the leader

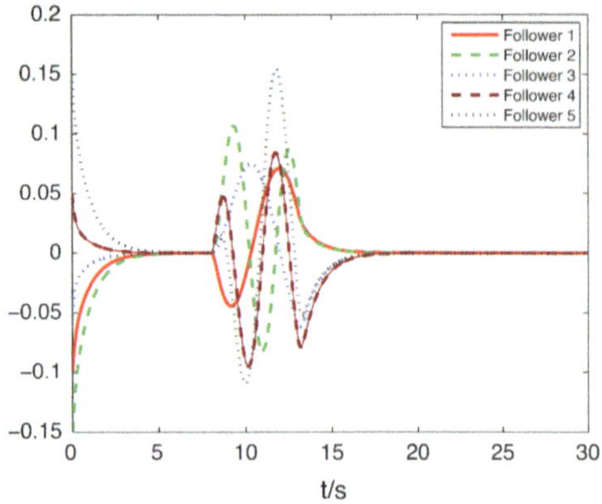

Fig. 7.6 Tracking errors

cally to small neighborhoods of the origin, respectively. However, if signal transmission faults occurred in the $8th$ second, i.e., $t = 8s$, then the tracking errors deviate from the neighborhood of the origin, which implies that the synchronization is no longer maintained. By using the FTC strategy (7.22), once the faults are compensated, it is easy to find, the synchronizing is obtained again. Simulation results show the effectiveness of the strategy proposed in this work.

7.5 Conclusions

In this chapter, the fault-tolerant synchronizing control problem is investigated for a class of uncertain nonlinear leader-following systems with communication faults. By using adaptive control and neural networks, an FTC strategy is designed, which ensures that the synchronization between the leader and followers is achieved regardless of the occurrence of the communication faults.

References

Bertsekas DP, Tsitsiklis JN (2007) Comments on "Coordination of groups of mobile autonomous agents using nearest neighbor rules." IEEE Trans Autom Control 52(5):968–969

Deng C, Yang G (2016) Distributed adaptive fault-tolerant containment control for a class of multi-agent systems with non-identical matching non-linear functions. IET Control Theory Appl 10(3):273–281

Deng C, Yang G (2017) Cooperative adaptive output regulation for linear multi-agent systems with actuator faults. IET Control Theory Appl 11(14):2396–2402

Deng C, Yang G (2018) Distributed adaptive fuzzy control for nonlinear multi-agent systems under directed graphs. IEEE Trans Fuzzy Syst 26(3):1356–1366

Ge SS, Hong F, Lee TH (2004) Adaptive neural control of nonlinear time-delay systems with unknown virtual control coefficients. IEEE Trans Syst Man Cybernet Part B 34(1):499–516

Gong J, Jiang B, Shen QK (2019) Distributed-observer-based faults tolerant control design for nonlinear multi-agent systems. Int J Control Autom Syst 17(12):3149–3157

Hua Y, Dong X, Li Q (2018) Distributed fault-tolerant time-varying formation control for second-order multi-agent systems with actuator failures and directed topologies. IEEE Trans Circuits Syst II Express Briefs 65(6):774–778

Jadbabaie A, Lin J, Morse AS (2003) Coordination of groups of mobile autonomous agents using nearest neighbor rules. IEEE Trans Autom Control 48(6):988–1001

Jiao H, Shi M, Shen QK (2018) Filter design with adaptation to time-delay parameters for genetic regulatory networks. IEEE/ACM Trans Comput Biol Bioinf 15(1):323–329

Jiao H, Zhang L, Shen QK (2018) Robust gene circuit control design for time-delayed genetic regulatory networks without SUM regulatory logic. IEEE/ACM Trans Comput Biol Bioinf 15(6):2086–2093

Jin X, Wang S, Qin J (2018) Adaptive fault-tolerant consensus for a class of uncertain nonlinear second-order multi-agent systems with circuit implementation. IEEE Trans Circuits Syst I Regul Pap 65(7):2243–2255

Ma Z, Ma H (2018) Adaptive finite-time dynamic output-feedback FTC design for MIMO nonlinear systems with actuator and sensor faults. IEEE Access 6:44824–44834

Ma H, Yang G (2016) Adaptive fault tolerant control of cooperative heterogeneous systems with actuator faults and unreliable interconnections. IEEE Trans Autom Control 61(11):3240–3255

Moreau L (2005) Stability of multiagent systems with time-dependent communication links. IEEE Trans Autom Control 50(2):169–182

Ren W, Cao C (2011) Distributed coordination of multi-agent networks: emergent problems, models, and issues. Springer, London

Ren W, Beard R, Atkins E (2007) Information consensus in multivehicle cooperative control. IEEE Control Syst Mag 27(2):71–82

Semsar E, Khorasani K (2008) Optimal consensus algorithms for cooperative team of agents subject to partial information. Automatica 44(11):2766–2777

Shen QK, Shi P (2016) Output consensus control of multiagent systems with unknown nonlinear Dead Zone. IEEE Trans Syst Man Cybernet: Syst 46(10):1329–1337

Shen QK, Jiang B, Shi P (2014) Cooperative adaptive fuzzy tracking control for networked unknown nonlinear multiagent systems with time-varying actuator Faults. IEEE Trans Fuzzy Syst 22(3):494–504

Shen QK, Shi P, Shi Y (2017) Adaptive output consensus with saturation and dead-zone and its application. IEEE Trans Industr Electron 64(6):5025–5034

Shen QK, Shi P, Zhu J (2019) Adaptive consensus control of leader-following systems with transmission nonlinearities. Int J Control 92(2):317–328

Shen QK, Shi P, Zhu J, Wang S, Shi Y (2020) Neural networks based distributed adaptive control of nonlinear multi-agent systems. IEEE Trans Neural Netw Learn Syst 31(3):1010–1021

Shen QK, Shi P, Agarwa Ramesh K (2021) Adaptive neural networks-based filter design for nonlinear systems with multiple constrains. IEEE Trans Neural Netw Learn Syst 32(7):3256–3261

Shi P, Shen QK (2017) Observer-based leader-following consensus of uncertain nonlinear multi-agent systems. Int J Robust Nonlinear Control 27(17):3794–3811

Xu X, Shen QK (2022) Fault detection and isolation observer design for a class of high-order dynamic systems. ICIC Express Lett 17(9):1–8

Ye D, Zhao X, Cao B (2016) Distributed adaptive fault-tolerant consensus tracking of multi-agent systems against time-varying actuator faults. IET Control Theory Appl 10(5):554–563

Ye D, Chen M, Yang H (2019) Distributed adaptive event-triggered fault-tolerant consensus of multiagent systems with general linear dynamics. IEEE Trans Cybernet 49(3):757–767

Zhang H, Lewis FL (2012) Adaptive cooperative tracking control of higher-order nonlinear systems with unknown dynamics. Automatica 48(7):1432–1439

Zhao L, Yu J, Lin C (2018) Adaptive neural consensus tracking for nonlinear multiagent systems using finite-time command filtered backstepping. IEEE Trans Syst Man Cybernet: Syst 48(11):2003–2012

Zheng X, Shen QK (2023) Adaptive fault-tolerant control for high-order nonlinear systems with supervisory controllers and command filters. Int J Control, Autom Syst 21(1):12–19

Zhu J, Shen QK, Zhang TP (2023) Neural network-based decentralized adaptive fault-tolerant control for a class of nonlinear interconnected systems with unknown input powers. Int J Adapt Control Signal Process 37(7):1693–1715

Zuo Z, Zhang J, Wang Y (2015) Adaptive fault tolerant tracking control for linear and lipschitz nonlinear multi-agent systems. IEEE Trans Industr Electron 62(2):3923–3931

Chapter 8
Filter-Based Distributed Fault-Tolerant Neural Control Against Sensor Faults

8.1 Introduction

During the past decades, Defoort et al. (2008), Yoo and Kim (2015), Wang et al. (2007), Fax and Murray (2003), Pan et al. (2015), multi-agent systems (MASs) including the leader-following systems (LFSs) considered in this chapter are often used to mode the complex systems, for example, power systems, multiple unmanned vehicles, mobile robots, and so on. For the research on MASs, due to its widespread application, the issue of distributed cooperative control has received increasing attention. For the leader-following systems, consensus or synchronization control is fundamental, which has been extensively investigated by many experts and scholars. For consensus or synchronization, there are generally two types: leaderless and leader-following consensus or synchronization. A large number of relevant results have been reported in literature Das and Lewis (2010), Zhang and Lewis (2012).

It is necessary to point out that, for nonlinear systems including uncertain and certain systems, adaptive control is an important control technique. Combining with neural networks (NNs) and fuzzy logic systems (FLSs), it often is applied in the modern controlled systems. In recent years, with the increasing application of multi-agent systems and/or leader-following systems, adaptive control technology has been applied to the distributed control of the systems, and A large number of adaptive control methods are proposed Cheng et al. (2010), Shen and Shi (2015), Peng et al. (2014), Wang and Tong (2018), Li et al. (2019). However, it is under healthy or fault-free condition that the above results were obtained. In fact, system components such as actuators and sensors fail during actual operation. To make matters worse, these failures can lead to the degradation of system performance and even worse accidents. Since the occurrence of faults is inevitable, these faults must be considered in control design. Hence, fault-tolerant control (FTC) is proposed and continuously improved. Fault-tolerant control, reduces or weakens the negative influence of the faults on system performance through fault compensation, and maintains system performance at an acceptable level or range.

Q. Shen, *Distributed Fault-Tolerant Consensus Control of Leader-Following Systems*, Intelligent Control and Learning Systems 11,
https://doi.org/10.1007/978-981-99-7426-9_8

Fault-tolerant control has made great progress and achieved fruitful results Shen et al. (2012; 2014), Xiao et al. (2012), Zhai et al. (2016). However, fault-tolerant control is mainly for simple single-agent systems. What's worse, fault-tolerant control methods designed for single-agent systems cannot be directly applied to MASs including the LFSs considered in this chapter. In recent years, the fault-tolerant control of multi-agent systems has received more and more attention Shen et al. (2013), Zhang et al. (2019). Compared with single-agent systems, MASs are more complex. For the MASs and LFSs, their FTC should be further studied although some preliminary results have been achieved.

As stated previously, radial basis function (RBF) neural networks, as a universal approximation, have been used widely in modern advanced control Cheng et al. (2010), Shen and Shi (2015), Peng et al. (2014). From the structure of the RBF NNs, it can be found that their input contains the system information such as system states. Once the states are polluted because of sensor faults and the polluted states are input to neural network, the neural network cannot accurately approximate an unknown function. Hence, how to compensate for these sensor faults to improve the approximation ability of the neural network is very important.

In this chapter, the distributed FTC problem is considered for nonlinear output feedback leader-following systems with sensor faults. The following contributions are worthy to emphasize:

(1) The influence of state sensor faults, which affects the ability of the neural network to approximate, is considered, and an effective method is introduced to ensure the approximation capability of neural networks regardless of the sensor faults;

(2) The proposed control scheme combines filters design method with the neighborhood output estimation errors, and an NNs-based backstepping FTC method is proposed such that the synchronization between the leader and followers is achieved.

8.2 Preliminaries

8.2.1 Graph Theory

As stated in Chap. 3, in the sense of mathematics, MASs including LFSs investigated in this book can be represented by a graph. Obviously, it involves basic graph theory in the research of MASs. So, let us first introduce it in this subsection. For basic graph theory, it was already introduced in the previous chapter and cannot be re-introduced here. Its detailed information can be found in the previous chapter.

8.2.2 Problem Formulation

Consider the following leader-following systems, which consist of one leader and N followers. Each follower has the following form:

$$\begin{cases} \dot{v}_{i,1} = v_{i,2} + f_{i,1}(y_i) + d_{i,1}(t) \\ \vdots \\ \dot{v}_{i,\rho-1} = v_{i,\rho} + f_{i,\rho-1}(y_i) + d_{i,\rho-1}(t) \\ \dot{v}_{i,\rho} = v_{i,\rho+1} + f_{i,\rho}(y_i) + u_i(t) + d_{i,\rho}(t) \\ \vdots \\ \dot{v}_{i,n-1} = v_{i,n} + f_{i,n-1}(y_i) + u_i(t) + d_{i,n-1}(t) \\ \dot{v}_{i,n} = f_{i,n}(y_i) + u_i(t) + d_{i,n}(t) \\ y_i = v_{i,1} \end{cases} \tag{8.1}$$

where $i = 1, \ldots, N$, state vector $v_i = [v_{i,1}, \ldots, v_{i,n}] \in R^n$, $v_{i,1} \in R$ $(l = 1, 2, \ldots, n)$, $u_i \in R$ and $y_i \in R$ denote respectively the state, input and output of the ith follower; function $f_{i,p}(y_i) \in R$ is smooth and unknown, and disturbance $d_{i,p}(t) \in R$ is unknown but bounded, $p = 1, \ldots, n$.

For the leader, which is labeled 0, has the following form:

$$\begin{cases} \dot{v}_{0,q} = \dot{v}_{0,q+1}, \quad q = 1, \ldots, n-1 \\ \dot{v}_{0,n} = f_0(y_0, t) \\ y_0 = v_{0,1} \end{cases} \tag{8.2}$$

where state vector $v_0 = [x_{0,1}, \ldots, v_{0,n}]^T \in R^n, v_{0,l} \in R\,(l = 1, 2, \ldots, n)$ and $y_0 \in R$ denote respectively the lth state and output of the leader; function $f_0(\cdot) \in R$. Note that, for all the followers, $f_0(\cdot)$ is assumed to be unknown.

From (8.1), let

$$A_i - \begin{bmatrix} 0 & I_{n-1} \\ 0 & 0 \end{bmatrix}, v_l = [v_{l,1}, v_{i,2}, \ldots, v_{i,n}]^T \in R^n,$$

$$B_{ui} = [0_{1\times(\rho-1)}, 1_{1\times(n-\rho+1)}]^T,$$

$$f_i(y_i) = \begin{bmatrix} f_{i,1}(y_i) \\ \vdots \\ f_{i,n}(y_i) \end{bmatrix}, \quad d_i = \begin{bmatrix} d_{l,1} \\ \vdots \\ d_{i,n} \end{bmatrix}, \quad C_i = [1, 0, \ldots, 0]$$

Then (8.1) is further presented as

$$\begin{cases} \dot{v}_i = A_i v_i + B_{ui} u_i + f_i(y_i) + d_i \\ y_i = C_i v x_i \end{cases} \tag{8.3}$$

In this work, the following sensor faults are considered,

$$y_{i,f} = y_i + b_{i,y} \tag{8.4}$$

where $b_{i,y} \in R$ denote the fault, which is unknown but bounded.

For follower i, the pth tracking error is defined as follows:

$$\epsilon_{i,p} = v_{i,p} - v_{0,p}, \quad i = 1, \ldots, N, \quad p = 1, \ldots, n \tag{8.5}$$

Define $\epsilon_p = [\epsilon_{1,p}, \ldots, \epsilon_{N,p}]^T \in R^N, \ p = 1, \ldots, n$, then one has

$$\epsilon_p = v_p - \bar{v}_{0,p} \tag{8.6}$$

where $\bar{v}_{0,p} = [v_{0,p}, \ldots, v_{0,p}]^T \in R^N$.

The main task in this chapter is, for the system (8.1), considering the sensor fault model (8.4), to construct an adaptive FTC method such that the synchronization between the leader and followers and the tracking error $\epsilon_{i,p}$ asymptotically converges to a small neighborhood of the origin, $p = 1, 2, \ldots, n, i = i, \ldots, N$.

Now, some assumptions are made, and a lemma is needed in this chapter.

Assumption 8.1 For the leader, its output and first derivative, i.e., y_0 and $\dot{y}_0$, are available and bounded.

Assumption 8.2 $d_{i,q}(t)$ is bounded, namely, $|d_{i,q}(t)| \leq \bar{d}_{i,q}$ with $\bar{d}_{i,q} > 0 \in R$ being a constant, $q = 1, \ldots, n, i = 1, \ldots, N$.

Assumption 8.3 $b_{i,y}$ is bounded, namely, $|b_{i,y}| \leq b_{i,0}$ with $b_{i,0} > 0 \in R$ being an unknown constant $i = 1, \ldots, N$.

Lemma 8.1 (Zhang and Lewis 2012) *$||\epsilon_q|| \leq ||z_q||/\underline{\varsigma}(B + L)$ with $\underline{\varsigma}(B + L)$ being the minimum singular value of matrix $B + L$, $q = 1, \ldots, n$.*

8.2.3 Neural Networks

In this work, RBF NNs Xia et al. (2016) are applied to approximate an unknown continuous function, which has been in the previous chapters. The detailed description of RBF NNs is omitted here. Interesting readers can find it from the previous chapters.

For a continuous function $h_i(v_i)$, it can be obtained

$$h_i(v_i) = \theta_i^{*T} \tau_i(\bar{v}_i) + \delta_i(\bar{v}_i) \tag{8.7}$$

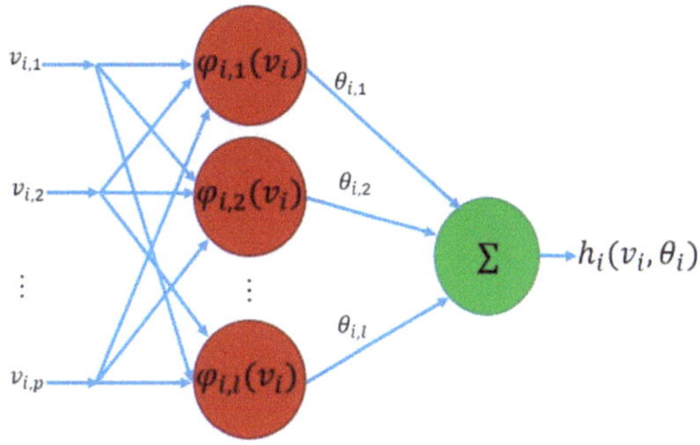

Fig. 8.1 RBF Neural Networks

where $\delta_i(\bar{v}_i)$ and θ_i^* represent the optimal approximation error and ideal weight vector, $\bar{v}_i = [v_i^T, 1]^T$.

Assumption 8.4 $\delta_i(\bar{v}_i)$ and θ_i^* are bounded, namely, $|\delta_i| \le \bar{\delta}_i$ with $\bar{\delta}_i \in R$ being an unknown constant.

From Fig. 8.1, for the approximation capability of RBF NNs, the exact values of the states play a very important role. As stated in the previous chapter, the values of the states are polluted due to the state sensor faults, which reduces the approximation capability of the neural networks.

Once the sensors become faulty, the obtained value of the state is described as $v_{i,f} = v_{i,f} + b_i$, where $v_{i,f} = [v_{i,f,1}, \ldots, v_{i,f,p}]^T \in R^p$, $b_i = [b_{i,1}, \ldots, b_{i,p}]^T \in R^p$. From the definition of $\tau_i(v_i)$ in the previous chapters, we have

$$\tau_i(\bar{v}_i) = [\tau_{i,1}(v_{i,1}),\ \tau_{i,2}(v_{i,2}), \ldots, \tau_{i,p}(v_{i,p})] \tag{8.8}$$

where $i = 1, \ldots, l$, $j = 1, \ldots, p$,

$$\tau_{i,j}(v_{i,j}) = \exp(-\frac{(v_{i,j} - \mu_{ij})^2}{(d_i)^2}). \tag{8.9}$$

From the mean value theorem, one has

$$\begin{aligned} \tau_{i,1}(v_{i,f,1}) &= \tau_i(v_{i,1}) + (-\frac{2(v_{i,f,1} - \mu_{i1})}{d_i^2})b_{i,1}\exp(-\frac{(v_{i,f,1} - \mu_{i1})^2}{d_i^2}) \\ &= \tau_{i,j}(v_{i,1}) + b_{i1\tau} \end{aligned} \tag{8.10}$$

Similarly, we have

$$\begin{aligned} \tau_{i,2}(v_{i,f,2}) &= \tau_{i,2}(v_{i,2}) + b_{i2\tau} \\ &\vdots \\ \tau_{i,p}(v_{i,f,p}) &= \tau_{i,p}(v_{i,p}) + b_{ip\tau} \end{aligned} \tag{8.11}$$

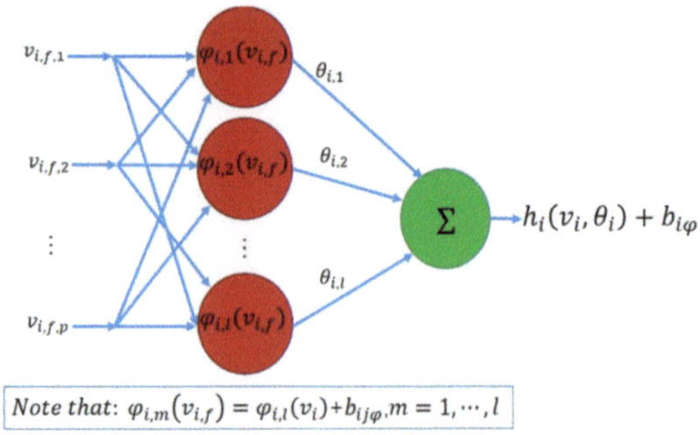

Fig. 8.2 RBF Neural Networks (faulty case)

where $i = 1, \ldots, l$,

$$b_{ij\tau} = (-\frac{2(v_{i,f,1} - \mu_{i1})}{(d_i)^2})b_1 \exp(-\frac{(v_{i,1,\tau} - \mu_{i1})^2}{(d_i)^2}) \in R \tag{8.12}$$

with $j = 1, \ldots, p$ are bounded. Then

$$\begin{aligned} \tau_{i,j}(v_{i,f}) &= (\tau_{i,j}(v_{i,1}) + b_{i1\tau})(\tau_{i,2}(v_{i,2}) + b_{i2\tau}) \times \cdots \times (\tau_{i,p}(v_{i,p}) + b_{ip\tau}) \\ &= \tau_{i,j}(v_i) + \bar{b}_{i\tau} \end{aligned} \tag{8.13}$$

where

$$\bar{b}_{i\tau} = \tau_{i,j}(v_{i,f}) - \tau_{i,j}(v_i) \tag{8.14}$$

which $\bar{b}_{i\tau}$ is bounded real number, $i = 1, \ldots, l$.

Then, from (8.4), $h_i(v_{i,f}, \theta)$ is presented as

$$\begin{aligned} h_i(v_{i,f}, \theta_i) &= \theta_i^T \tau_i(v_{i,f}) \\ &= \theta_i^T [\tau_i(v_i) + b_{i\tau}] \\ &= \theta_i^T \tau_i(v_i) + \theta_{i\tau} \end{aligned} \tag{8.15}$$

where $b_{i\tau} = [b_{i1\tau}, \ldots, b_{il\tau}]$, $\theta_{i\tau} = \theta^T b_{i\tau}$ is bounded.

Under the faulty condition, the structure of RBF NNs is shown in Fig. 8.2.

8.3 Main Results

8.3.1 Filter Design

In this chapter, the unknown functions in (8.1) are approximated by RBF NNs. Because of the state sensor faults, based on the analysis in the previous subsection, one has

$$\tau(y_{i,f}) = \tau(y_i) + \omega_i \tag{8.16}$$

where vector ω_i is bounded. From (8.1), it follows

$$\begin{aligned} f_{i,1}(y_i) &= [\tau_{i,1}(y_{i,f}) - \omega_{i,1}]^T \theta_{i,1}^* + \delta_{i,1}(y_i) \\ &= \tau_{i,1}^T(y_{i,f})\theta_{i,1}^* - \theta_{i,\omega 1} + \delta_{i,1}(y_i) \\ &\vdots \\ f_{i,n}(y_i) &= \tau_{i,n}^T(y_{i,f})\theta_{i,n}^* - \theta_{i,wn} + \delta_{i,n}(y_i) \end{aligned} \tag{8.17}$$

where $\theta_{i,\omega q} = \omega_{i,q}^T \theta_{i,q}^* \in R, q = 1, \ldots, n, i = 1, \ldots, N$.

Note that, for $f_{i,q}(y_i)$ $(i = 1, \ldots, N, q = 1, \ldots, n)$, $\tau_{i,q}^T, \theta_{i,q}^*$ and $\delta_{i,q}$ are bounded, thus $\theta_{i,\omega q}$ is also bounded.

From (8.3) and (8.17), it follows

$$\begin{cases} \dot{x}_i = A_i x_i + B_{ui} u_i + \tau_i^T(y_{i,f})\theta_i^* - \theta_{i,w} + \delta_i + d_i \\ y_i = C_i x_i \end{cases} \tag{8.18}$$

where

$$\tau_i^T(y_{i,f}) = \begin{bmatrix} \tau_{i,1}^T & & \\ & \ddots & \\ & & \tau_{i,n}^T \end{bmatrix}, \theta_i^* = \begin{bmatrix} \theta_{i,1}^* \\ \vdots \\ \theta_{i,n}^* \end{bmatrix},$$

$$\theta_{i,\omega} = \begin{bmatrix} \theta_{i,\omega 1} \\ \vdots \\ \theta_{i,wn} \end{bmatrix}, \delta_i = \begin{bmatrix} \delta_{i,1} \\ \vdots \\ \delta_{i,n} \end{bmatrix}, d_i = \begin{bmatrix} d_{i,1} \\ \vdots \\ d_{i,n} \end{bmatrix}$$

For follower $i(i = 1, \ldots, N)$, design the following filter:

$$\dot{\xi}_i - A_{i,0}\xi_i + k_i y_{i,f} + k_i \sum_{j \in N_i} a_{ij}[(y_{i,f} - \hat{y}_{i,f}) - (y_{j,f} - \hat{y}_{j,f})], \quad \xi_i \in R^n \tag{8.19}$$

$$\dot{\vartheta}_i = A_{i,0}\vartheta_i + \tau_i^T(y_{i,f}), \vartheta_i \in R^{n \times \bar{l}} \tag{8.20}$$

$$\dot{\lambda}_i = A_{i,0}\lambda_i + e_{i,n}u_i, \lambda_i \subset R^n \tag{8.21}$$

$$w_{i,p} = A_{i,0}^p \lambda_i, p = 0, 1, \ldots, m \tag{8.22}$$

where $\hat{y}_{i,f}$ denotes the estimate of $y_{i,f}$, $\bar{l} = l_1 + \cdots + l_n$, $m = n - \rho$. Observer gain $k_i = [k_{i,1}, \ldots, k_{i,n}]^T$ is chosen appropriately to ensure $A_{i,0} = A_i - k_i C_i, i = 1, \ldots, N$ is a Hurwitz matrix.

From (8.19), we have

$$\begin{aligned}\dot{\xi}_i &= A_{i,0}\xi_i + k_i y_{i,f} + k_i \sum_{j\in N_i} g_{ij}[(y_{i,f} - \hat{y}_{i,f}) - (y_{j,f} - \hat{y}_{j,f})] \\ &= A_{i,0}\xi_i + k_i y_{i,f} + k_i[g_{i1}((y_{i,f} - \hat{y}_{i,f}) - (y_{1,f} - \\ &\quad \hat{y}_{1,f})) + \cdots + g_{iN_i}((y_{i,f} - \hat{y}_{i,f}) - (y_{N_i,f} - \hat{y}_{N_i,f}))] \\ &= A_{i,0}\xi_i + k_i y_{i,f} + (g_{i1} + \cdots + g_{iN_i})(y_{i,f} - \hat{y}_{i,f})k_i - \\ &\quad g_{i1}(y_{1,f} - \hat{y}_{1,f})k_i - \cdots - g_{iN_i}(y_{N_i,f} - \hat{y}_{N_i,f})k_i\end{aligned} \tag{8.23}$$

Further, (8.23) is presented in the following compact form,

$$\dot{\xi} = A_0\xi + KL(y_f - \hat{y}_f) + Ky_f, \quad \xi \in R^{nN} \tag{8.24}$$

where $y_f = [y_{1,f}, \ldots, y_{N,f}]^T \in R^N$, $\hat{y}_f = [\hat{y}_{1,f}, \ldots, \hat{y}_{N,f}]^T \in R^N$ is the estimate of y_f, $A_0 = diag\{A_{1,0}, \ldots, A_{N,0}\} \in R^{nN\times nN}$, $\xi = [\xi_1, \ldots, \xi_N]^T \in R^{nN}$, $K = diag\{k_1, \ldots, k_N\} \in R^{nN\times N}$.

Thus, (8.20)–(8.22) are presented, respectively, in the following forms:

$$\dot{\vartheta} = A_0\vartheta + \tau^T(y_f), \vartheta \in R^{nN\times l\bar{N}} \tag{8.25}$$

$$\dot{\lambda} = A_0\lambda + e_n u, \lambda \in R^{nN} \tag{8.26}$$

$$w_p = A_0^p\lambda, \; p = 0, 1, \ldots, m \tag{8.27}$$

where $\vartheta = diag\{\vartheta_1, \ldots, \vartheta_N\} \in R^{nN\times l\bar{N}}$, $\tau^T(y_f) = diag\{\tau_1^T(y_{1,f}), \ldots, \tau_N^T(y_{N,f})\} \in R^{nN\times l\bar{N}}$, $\lambda = [\lambda_1, \ldots, \lambda_N]^T \in R^{nN}$, $e_n = diag\{e_{1,n}, \ldots, e_{N,n}\} \in R^{nN\times N}$, $w_q = [w_{1,q}, \ldots, w_{N,q}]^T \in R^{nN}$, $u = [u_1, \ldots, u_N]^T \in R^N$.

From (8.18), (8.1) is re-described

$$\begin{cases}\dot{v} = Av + B_u u + \tau^T(y_f)\theta^* - \theta_\omega + d + \delta \\ y_f = Cx\end{cases} \tag{8.28}$$

where $A = diag\{A_1, \ldots, A_N\} \in R^{nN\times nN}$, $x = [x_1, \ldots, x_N]^T \in R^{nN}$, $B_u = diag\{B_{u1}, \ldots, B_{uN}\} \in R^{nN\times N}$, $\theta_\omega = [\theta_{1,\omega}, \ldots, \theta_{N,\omega}]^T \in R^{nN}$, $\theta^* = [\theta_1^*, \ldots, \theta_N^*]^T \in R^{nN}$, $\delta = [\delta_1, \ldots, \delta_N]^T \in R^{nN}$, $d = [d_1, \ldots, d_N]^T \in R^{nN}$, $y_f = [y_{1,f}, \ldots, y_{N,f}]^T \in R^{nN}$, $C = diag\{C_1, \ldots, C_N\} \in R^{N\times nN}$.

For follower i, design the observer

$$\hat{v}_i = \xi_i + \vartheta_i\hat{\theta}_i + \sum_{p=0}^{m} w_{i,p} \tag{8.29}$$

where $\hat{v}_i = [\hat{v}_{i,1}, \ldots, \hat{v}_{i,n}]^T \in R^n$ is the estimate of v_i, $\hat{\theta}_i = [\hat{\theta}_{i,1}, \ldots, \hat{\theta}_{i,n}]^T \in R^{\bar{l}}$ is the estimate of θ_i^*.

Further, one has

$$\hat{v} = \xi + \vartheta\hat{\theta} + \sum_{p=0}^{m} w_p \tag{8.30}$$

where $\hat{v} = [\hat{v}_1^T, \ldots, \hat{v}_N^T]^T \in R^{nN}$, $\hat{\theta} = [\hat{\theta}_1, \ldots, \hat{\theta}_N]^T \in R^{l\bar{N}}$.

Let observer error $\varepsilon = v - \hat{v}$, then one has

$$\begin{aligned}\dot{\varepsilon} &= \dot{v} - \dot{\hat{v}} \\ &= Ax + Bu + \tau^T(y_f)\theta^* - \theta_\omega + \delta + d - A_0\xi - \\ &\quad Ky_f - KL(\tilde{y} + \tilde{b}_y) - [A_0\vartheta + \tau^T(y_f)]\hat{\theta} - \vartheta\dot{\hat{\theta}} - \\ &\quad \sum_{p=0}^{m} A_0^p(A_0\lambda + e_n u) \\ &= (A_0 - KLC)\varepsilon - \vartheta\dot{\hat{\theta}} + \tau^T(y_f)\tilde{\theta} - KL\tilde{b}_y - \theta_\omega + d + \delta\end{aligned} \tag{8.31}$$

where $\tilde{b}_y = b_y - \hat{b}_y$, $b_y = [b_{1,y}, \ldots, b_{N,y}]^T \in R^N$ with $\hat{b}_y$ being the estimate of b_y, $\tilde{y} = y - \hat{y}$.

Then, matrices $P = P^T > 0$ and K are chosen appropriately to ensure the following inequality (8.32) is satisfied,

$$(A_0 - KDC)^T P + P(A_0 - KDC) \leq -hI \tag{8.32}$$

where $P = diag\{P_1, \ldots, P_N\} \in R^{nN \times nN}$, $h > 0 \in R$.

Define

$$V_\varepsilon = \varepsilon^T P\varepsilon \tag{8.33}$$

From (8.31), its time derivative of V_ε is described as

$$\begin{aligned}\dot{V}_\varepsilon &= [\tau^T(y_f)\tilde{\theta} + (A_0 - KLC)\varepsilon - KL\tilde{b}_y - \theta_\omega + \delta - \\ &\quad \vartheta\dot{\hat{\theta}} + d]^T P\varepsilon + \varepsilon^T P[\tau^T(y_f)\tilde{\theta} + (A_0 - KLC)\varepsilon - \\ &\quad KL\tilde{b}_y - \theta_\omega - \vartheta\dot{\hat{\theta}} + d + \delta] \\ &= [\tau^T(y_f)\tilde{\theta} + (A_0 - K(D-G)C)\varepsilon - K(D- \\ &\quad G)\tilde{b}_y - \theta_\omega - \vartheta\dot{\hat{\theta}} + d + \delta]^T P\varepsilon + \varepsilon^T P[(A_0 - \\ &\quad K(D-G)C)\varepsilon - K(D-G)\tilde{b}_y + \tau^T(y_f)\tilde{\theta} - \\ &\quad \theta_\omega - \vartheta\dot{\hat{\theta}} + d + \delta] \\ &= \varepsilon^T[(A_0 - KDC)^T P + P(A_0 - KDC)]\varepsilon + \\ &\quad 2\varepsilon^T P\tau^T(y_f)\tilde{\theta} + 2\varepsilon^T PKG\tilde{b}_y + 2\varepsilon^T PKGC\varepsilon - \\ &\quad 2\varepsilon^T P\theta_\omega - 2\varepsilon^T P\vartheta\dot{\hat{\theta}} - 2\varepsilon^T PKD\tilde{b}_y + 2\varepsilon^T Pd + 2\varepsilon^T P\delta\end{aligned} \tag{8.34}$$

The $4th$ and $5th$ terms in (8.34) can be developed as

$$\begin{aligned}
&2\varepsilon^T P\tau^T(y_f)\tilde{\theta}\\
&=2[\varepsilon_1 P_1+\cdots+\varepsilon_N P_N]\begin{bmatrix}\tau_1^T\tilde{\theta}_1\\ \vdots\\ \tau_N^T\tilde{\theta}_N\end{bmatrix}\\
&=2(\varepsilon_1 P_1\tau_1^T\tilde{\theta}_1+\cdots+\varepsilon_N P_N\tau_N^T\tilde{\theta}_N)\\
&=2[\varepsilon_{1,1}M_{1,p}^T\tau_1^T\tilde{\theta}_1+(\varepsilon_{1,2},\cdots+\varepsilon_{1,n})\bar{P}_1\tau_1^T\tilde{\theta}_1+\cdots+\\
&\varepsilon_{N,1}M_{N,p}^T\tau_N^T\tilde{\theta}_N+\cdots+(\varepsilon_{N,2},\ldots,\varepsilon_{N,n})\bar{P}_N\tau_N^T\tilde{\theta}_N]\\
&=2M_\varepsilon^T M_p\tau^T\tilde{\theta}+2\bar{\varepsilon}^T\bar{P}\tau^T\tilde{\theta}
\end{aligned}\tag{8.35}$$

and

$$\begin{aligned}
&-2\varepsilon^T PKD\tilde{b}_y\\
&=-2[\varepsilon_1 P_1 k_1+\cdots+\varepsilon_N P_N k_N]\begin{bmatrix}d_1\tilde{b}_{1,y}\\ \vdots\\ d_N\tilde{b}_{N,y}\end{bmatrix}\\
&=-2(\varepsilon_1 P_1 k_1 d_1\tilde{b}_{1,y}+\cdots+\varepsilon_N P_N k_N d_N\tilde{b}_{N,y})\\
&=-2[\varepsilon_{1,1}M_{1,p}^T k_1 d_1\tilde{b}_{1,y}+(\varepsilon_{1,2},\ldots,\varepsilon_{1,n})\bar{P}_1 k_1 d_1\tilde{b}_{1,y}+\\
&\varepsilon_{N,1}M_{N,p}^T k_N d_N\tilde{b}_{N,y}+\cdots+(\varepsilon_{N,2},\ldots,\varepsilon_{N,n})\times\\
&\bar{P}_N k_N d_N\tilde{b}_{N,y}]\\
&=-2M_\varepsilon^T M_p KD\tilde{b}_y-2\bar{\varepsilon}^T\bar{P}KD\tilde{b}_y
\end{aligned}\tag{8.36}$$

where

$$M_\varepsilon=[\varepsilon_{1,1},\varepsilon_{2,1},\ldots,\varepsilon_{N,1}]^T\in R^N,\ \ \bar{\varepsilon}=[\varepsilon_{1,2},\ldots,\varepsilon_{1,n},\ldots,\varepsilon_{N,2},\ldots,\varepsilon_{N,n}]^T\in R^{(n-1)N},$$
$$M_p=diag\{M_{1,p}^T,\ldots,M_{N,p}^T\}\in R^{N\times nN},\ \ M_{i,p}=[P_{i,11},P_{i,12},\ldots,P_{i,1n}]^T\in R^n,$$
$$\bar{P}=diag\{\bar{P}_1,\ldots,\bar{P}_N\}\in R^{(n-1)N\times nN},\ \ \bar{P}_i=\begin{bmatrix}P_{i,21}&\cdots&P_{i,2n}\\ \vdots&\ddots&\vdots\\ P_{i,n1}&\cdots&P_{i,nn}\end{bmatrix}\in R^{(n-1)\times n}, i=1,\ldots,N$$

Substituting (8.35) and (8.36) into (8.34), it yields

$$\begin{aligned}
\dot{V}_\varepsilon=&\varepsilon^T[P(A_0-KDC)+(A_0-KDC)^T P]\varepsilon+\\
&2\varepsilon^T PKG\tilde{b}_y+2\varepsilon^T PKGC\varepsilon+2M_\varepsilon^T M_p\tau^T\tilde{\theta}+\\
&2\bar{\varepsilon}^T\bar{P}\tau^T\tilde{\theta}-2M_\varepsilon^T M_p KD\tilde{b}_y-2\bar{\varepsilon}^T\bar{P}KD\tilde{b}_y+\\
&2\varepsilon^T Pd+2\varepsilon^T P\delta-2\varepsilon^T P\theta_\omega-2\varepsilon^T P\vartheta\dot{\hat{\theta}}
\end{aligned}\tag{8.37}$$

From Young's Inequality and Assumptions 8.2–8.4, it follows

$$-2\varepsilon^T P\theta_\omega \le r\varepsilon^T\varepsilon + \frac{||P||^2||\theta_\omega||^2}{r} \tag{8.38}$$

$$2\varepsilon^T P\delta \le r\varepsilon^T\varepsilon + \frac{1}{r}||\delta||^2||P||^2 \le r\varepsilon^T\varepsilon + \frac{1}{r}||\bar{\delta}||^2||P||^2 \tag{8.39}$$

$$2\varepsilon^T Pd \le r\varepsilon^T\varepsilon + \frac{1}{r}||d||^2||P||^2 \le r\varepsilon^T\varepsilon + \frac{1}{r}||\bar{d}||^2||P||^2 \tag{8.40}$$

$$-2\varepsilon^T P\vartheta\dot{\hat{\theta}} \le \frac{1}{r}(P\vartheta\dot{\hat{\theta}})^T P\vartheta\dot{\hat{\theta}} + r\varepsilon^T\varepsilon \tag{8.41}$$

$$2\bar{\varepsilon}^T\bar{P}\tau^T\tilde{\theta} \le \tilde{\theta}^T\tau\bar{P}^T\bar{P}\tau^T\tilde{\theta} + \bar{\varepsilon}^T\bar{\varepsilon} \le \tilde{\theta}^T\tau\bar{P}^T\bar{P}\tau^T\tilde{\theta} + \varepsilon^T\varepsilon \tag{8.42}$$

$$\begin{aligned} -2\bar{\varepsilon}^T\bar{P}KD\tilde{b}_y &\le \tilde{b}_y^T D^T K^T\bar{P}^T\bar{P}KD\tilde{b}_y + \bar{\varepsilon}^T\bar{\varepsilon} \\ &\le \tilde{b}_y^T D^T K^T\bar{P}^T\bar{P}KD\tilde{b}_y + \varepsilon^T\varepsilon \end{aligned} \tag{8.43}$$

$$2\varepsilon^T PKGC\varepsilon \le 2\bar{\varsigma}(G)\bar{\varsigma}(K)\bar{\varsigma}(P)\varepsilon^T\varepsilon \tag{8.44}$$

$$\begin{aligned} 2\varepsilon^T PKG\tilde{b}_y \le & \bar{\varsigma}(G)\bar{\varsigma}(K)\bar{\varsigma}(P)\tilde{b}_y^T\tilde{b}_y + \\ & \bar{\varsigma}(G)\bar{\varsigma}(K)\bar{\varsigma}(P)\varepsilon^T\varepsilon \end{aligned} \tag{8.45}$$

where design parameter $r > 0$, $\bar{\varsigma}(\star)$ denote the maximum singular value of matrix $\star$ with $\star = P, K, G$, $\bar{d} = [\bar{d}_1, \ldots, \bar{d}_N]^T$, $\bar{\delta} = [\bar{\delta}_1, \ldots, \bar{\delta}_N]^T$, $\bar{d}_i = [\bar{d}_{i,1}, \ldots, \bar{d}_{i,n}]^T$, $\bar{\delta}_i = [\bar{\delta}_{i,1}, \ldots, \bar{\delta}_{i,n}]^T$, $i = 1, \ldots, N$.

Let $\bar{\gamma} = [\gamma, \ldots, \gamma]^T \in R^N$, $\gamma = \frac{||P||^2}{r}(||\theta_\omega||^2 + ||\bar{d}||^2 + ||\bar{\delta}||^2)$.

Substituting (8.38)–(8.45) into (8.37), we have

$$\begin{aligned} \dot{V}_\varepsilon \le & -[h - (2 + 4r + 3\bar{\varsigma}(G)\bar{\varsigma}(K)\bar{\varsigma}(P))]\varepsilon^T\varepsilon + \\ & 2M_\varepsilon^T M_p\tau^T\tilde{\theta} - 2M_\varepsilon^T M_p KD\tilde{b}_y + \\ & \tilde{\theta}^T\tau\bar{P}^T\bar{P}\tau^T\tilde{\theta} + ||P||^2(||\theta_\omega||^2 + d^2 + ||\bar{\delta}||^2)/r + \\ & \bar{\varsigma}(P)\bar{\varsigma}(K)\bar{\varsigma}(G)\tilde{b}_y^T\tilde{b}_y + (\bar{P}\vartheta\dot{\hat{\theta}})^T P\vartheta\dot{\hat{\theta}}/r + \tilde{b}_y^T D^T K^T\bar{P}^T\bar{P}KD\tilde{b}_y \end{aligned} \tag{8.46}$$

From (8.22), for vectors $w_{i,p}$ and λ_i, $w_{i,pq}$ and $\lambda_{i,k}$ denote their qth and kth element, respectively. $w_{i,pq}$ satisfies Hang and Zhang (2002):

$$\begin{aligned} & w_{i,pq} = [*, \ldots, *, 1]\begin{bmatrix} \lambda_{i,1} \\ \vdots \\ \lambda_{i,p+q} \end{bmatrix}, \\ & q = 1, \ldots, n, \ \lambda_{i,k} = 0, k > n, \ p = 0, \ldots, m, \end{aligned} \tag{8.47}$$

From (8.18), $v_{i,2}$ becomes

$$\begin{aligned} v_{i,2} =& \xi_{i,2} + \vartheta_{i,(2)}\hat{\theta}_i + [w_{i,m2}, \ldots, w_{i,12}, w_{i,02}]I_m + \varepsilon_{i,2} \\ =& \xi_{i,2} + w_{i,m2} + \vartheta_{i,(2)}\hat{\theta}_i + \\ & [0, w_{i,(m-1)2}, \ldots, w_{i.02}]I_m + \varepsilon_{i,2} \end{aligned} \tag{8.48}$$

where $I_m = [1, \ldots, 1]^T \in R^{m+1}$, $\vartheta_{i,(2)}$ and $\xi_{i,2}$ represent respectively the $2nd$ row of $\vartheta_{i,(2)}$ and the $2nd$ element of ξ_i.

Let

$$\begin{aligned} &\bar{v}_2 = [v_{1,2}, \ldots, v_{N,2}]^T, \bar{\xi}_2 = [\xi_{1,2}, \ldots, \xi_{N,2}]^T, \\ &\bar{\varepsilon}_2 = [\varepsilon_{1,2}, \ldots, \varepsilon_{N,2}]^T, \bar{\vartheta}_{(2)} = diag\{\vartheta_{1,(2)}, \ldots, \vartheta_{N,(2)}\}, \\ &\hat{\theta} = [\hat{\theta}_1, \ldots, \hat{\theta}_N]^T, w_{m2} = [w_{1,m2}, \ldots, w_{N,m2}]^T \\ &\bar{\bar{w}} = \begin{bmatrix} 0 & \cdots & w_{1,02} \\ \vdots & \ddots & \vdots \\ 0 & \cdots & w_{N,02} \end{bmatrix}, \bar{w} = \begin{bmatrix} w_{1,m2} & \cdots & w_{1,02} \\ \vdots & \ddots & \vdots \\ w_{N,m2} & \cdots & w_{N,02} \end{bmatrix}. \end{aligned}$$

From (8.48), it follows

$$\begin{aligned} \bar{v}_2 =& \bar{\xi}_2 + \bar{\vartheta}_{(2)}\hat{\theta} + \bar{w}I_m + \bar{\varepsilon}_2 \\ =& \bar{\xi}_2 + w_{m2} + \bar{\vartheta}_{(2)}\hat{\theta} + \bar{\bar{w}}I_m + \bar{\varepsilon}_2 \end{aligned} \tag{8.49}$$

From (8.28) and (8.48), we have

$$\begin{aligned} \dot{y}_f =& \bar{\xi}_2 + w_{m2} + \bar{\vartheta}_{(2)}\hat{\theta} + \bar{\bar{w}}I_m + \bar{\varepsilon}_2 + \bar{\tau}_{(1)}^T(y_f)\theta^* + \\ & w_{\delta 1} - w_{\theta\omega 1} + w_{d1} \\ =& \bar{\xi}_2 + \bar{\vartheta}_{(2)}\hat{\theta} + \bar{w}I_m + \bar{\varepsilon}_2 + \bar{\tau}_{(1)}^T(y_f)\theta^* + w_{\delta 1} - w_{\theta\omega 1} + w_{d1} \end{aligned} \tag{8.50}$$

where $\bar{\tau}_{(1)}^T = diag\{\tau_{1,(1)}^T, \ldots, \tau_{N,(1)}^T\}$, $w_{\delta 1} = [\delta_{1,1}, \ldots, \delta_{N,1}]^T$, $w_{\theta\omega 1} = [\theta_{1,\omega 1}, \ldots, \theta_{N,\omega 1}]^T$, $w_{d1} = [d_{1,1}, \ldots, d_{N,1}]^T$.

From (8.24)–(8.27) and (8.50), (28) can be re-described as:

$$\begin{cases} \dot{y}_f =& \bar{\xi}_2 + w_{m2} + \bar{\vartheta}_{(2)}\hat{\theta} + \bar{\bar{w}}I_m + \bar{\varepsilon}_2 + \bar{\tau}_{(1)}^T(y_f)\theta^* + \\ & w_{\delta 1} - w_{\theta\omega 1} + w_{d1} \\ \dot{w}_{mq} =& w_{m(q+1)} - k_q w_{m1}, q = 2, \ldots, \rho - 1 \\ \dot{w}_{m\rho} =& u + w_{m(\rho+1)} - k_\rho w_{m1} \end{cases} \tag{8.51}$$

8.3.2 Fault-Tolerant Controller Design

In this subsection, combining backstepping control technology with the K-filters designed in the above subsection, an adaptive output-feedback FTC method is constructed.

Let

$$\begin{aligned} z_{i,1} &= \sum_{j\in N_i} a_{ij}(y_i - y_j) + \mu_i(y_i - y_0) \\ z_{i,q} &= w_{i,mq} - \alpha_{i,q-1}, q = 2, \ldots, \rho \end{aligned} \tag{8.52}$$

where weight $a_i \geq 0$, virtual control law $\alpha_{i,q-1}$, will be designed later.

Because sensors become faulty, $y_{i,f} = y_i + b_{i,y}$. Then, one has

$$\begin{aligned} z_{i,1} =& \sum_{j\in N_i} a_{ij}[(y_{i,f} - b_{i,y}) - (y_{j,f} - b_{j,y})] + \mu_i[(y_{i,f} - b_{i,y}) - y_0] \\ =& \sum_{j\in N_i} a_{ij}[(y_{i,f} - \hat{b}_{i,y}) - (y_{j,f} - \hat{b}_{j,y})] + \mu_i[(y_{i,f} - \\ & \hat{b}_{i,y}) - y_0] - \left[\sum_{j\in N_i} a_{ij}(\tilde{b}_{i,y} - \tilde{b}_{j,y}) + \mu_i\tilde{b}_{i,y}\right] \\ z_{i,q} =& w_{i,mq} - \alpha_{i,q-1}, q = 2, \ldots, \rho \end{aligned} \tag{8.53}$$

where $\hat{b}_{i,y}$ denotes the estimate of $b_{i,y}$, $\tilde{b}_{i,y} = b_{i,y} - \hat{b}_{i,y}, i = 1, \ldots, N$.

Let $z_1 = [z_{1,1}, \ldots, z_{N,1}]^T \in R^N$, $y = [y_1, \ldots, y_N]^T$, $y_f = [y_{1,f}, \ldots, y_{N,f}]^T$, $b_y = [b_{1,y}, \ldots, b_{N,y}]^T \in R^N$, $\bar{y}_0 = [y_0, \ldots, y_0]^T \in R^N$, $B = diag\{\mu_1, \ldots, \mu_N\}$, $z_q = [z_{1,q}, \ldots, z_{1,q}]^T$, $w_{mq} = [w_{1,mq}, \ldots, w_{N,mq}]^T$, $\alpha_{q-1} = [\alpha_{1,q-1}, \ldots, \alpha_{N,q-1}]^T$, $q = 2, \ldots, \rho$, from (8.53), one has

$$\begin{aligned} z_1 &= (L + B)(y - \bar{y}_0) \\ &= (L + B)(y_f - \hat{b}_y - \bar{y}_0) - (L + B)\tilde{b}_y \\ &= \bar{z}_1 - (L + B)\tilde{b}_y \\ z_q &= w_{mq} - \alpha_{q-1}, q = 2, \ldots, \rho \end{aligned} \tag{8.54}$$

here $\bar{z}_1 = (B + L)(y_f - \hat{b}_y - \bar{y}_0)$, which is applied in virtual control designing.

Step 1: From (8.54), one has

$$\begin{aligned} \dot{z}_1 &= (B + L)(\dot{y}_f - \dot{\bar{y}}_0) \\ &= (B + L)(\bar{\xi}_2 + w_{m2} + \bar{\vartheta}_{(2)}\hat{\theta} + \bar{\bar{w}} I_m + \bar{\varepsilon}_2 + \\ &\quad \bar{\tau}_{(1)}^T(y_f)\theta^* + w_{\delta 1} - w_{\theta\omega 1} + w_{d1} - \dot{\bar{y}}_0) \end{aligned} \tag{8.55}$$

Since $w_{m2} = z_2 + \alpha_1$, it yields

$$\begin{aligned}\dot{z}_1 =&(B+L)(\bar{\xi}_2 + z_2 + \alpha_1 + \bar{\vartheta}_{(2)}\hat{\theta} + \bar{\bar{w}} I_m + \\ &\bar{\varepsilon}_2 + \bar{\tau}_{(1)}^T \theta^* + w_{\delta 1} - w_{\theta\omega 1} + w_{d1} - \dot{\bar{y}}_0)\end{aligned} \tag{8.56}$$

Now, α_1 is defined as

$$\alpha_1 = -c_1(D+B)^{-1}\bar{z}_1 - \bar{\xi}_2 - \bar{\vartheta}_{(2)}\hat{\theta} - \bar{\bar{w}} I_m - \hat{\bar{\gamma}} - \bar{\tau}_{(1)}^T(y_f)\hat{\theta} + \dot{\bar{y}}_0 \tag{8.57}$$

where $\hat{\bar{\gamma}}$ and $\hat{\theta}$ respectively denote the estimate of $\bar{\gamma}$ and θ^*. design parameter $c_1 > 0$ satisfying

$$c_1 - \frac{3}{2}\bar{\varsigma}(G) - \left(\frac{1}{2}c_1 + l_1 + 1\right)\bar{\varsigma}(B+L) - \frac{c_1\bar{\varsigma}(G)(\bar{\varsigma}(B+L)+2)}{2\underline{\varsigma}(B+D)} > 0$$

Define

$$V_1 = V_\varepsilon + \frac{1}{2}z_1^T z_1 + \frac{1}{2\eta_\theta}\tilde{\theta}^T\tilde{\theta} + \frac{1}{2\eta_\gamma}\tilde{\bar{\gamma}}^T\tilde{\bar{\gamma}} + \frac{1}{\eta_b}\tilde{b}_y^T\tilde{b}_y \tag{8.58}$$

where design constants $\eta_\theta > 0$, $\eta_\gamma > 0$ and $\eta_b > 0$, $\tilde{b}_y = b_y - \hat{b}_y$, $\tilde{\bar{\gamma}} = \bar{\gamma} - \hat{\bar{\gamma}}$, $\tilde{\theta} = \theta^* - \hat{\theta}$.

From (8.46) and (8.56), one has

$$\begin{aligned}\dot{V}_1 \le& -[h - (2 + 4r + 3\bar{\varsigma}(G)\bar{\varsigma}(K)\bar{\varsigma}(P))]\varepsilon^T\varepsilon + \\ &2M_\varepsilon^T M_p \tau^T\tilde{\theta} - 2M_\varepsilon^T M_p K D\tilde{b}_y + \\ &||P||^2(||\theta_\omega||^2 + \bar{d}^2 + ||\delta||^2)/r + \tilde{\theta}^T\tau\bar{P}^T\bar{P}\tau^T\tilde{\theta} + \\ &(\bar{P}\vartheta\dot{\hat{\theta}})^T\bar{P}\vartheta\dot{\hat{\theta}}/r + \tilde{b}_y^T D^T K^T\bar{P}^T P\bar{K}D\tilde{b}_y + \\ &\bar{\varsigma}(G)\bar{\varsigma}(K)\bar{\varsigma}(P)\tilde{b}_y^T\tilde{b}_y + z_1^T[(B+L)(\bar{\xi}_2 + z_2 + \alpha_1 + \\ &\bar{\vartheta}_{(2)}\hat{\theta} + \bar{\bar{w}} I_m + \bar{\varepsilon}_2 + \bar{\tau}_{(1)}^T(y_f)\theta^* + w_{\delta 1} - w_{\theta\omega 1} + \\ &w_{d1} - \dot{\bar{y}}_0)] - \frac{1}{\eta_b}\tilde{b}_y^T\dot{\hat{b}}_y - \frac{1}{\eta_\gamma}\tilde{\bar{\gamma}}^T\dot{\hat{\bar{\gamma}}} - \frac{1}{\eta_\theta}\tilde{\theta}^T\dot{\hat{\theta}}\end{aligned} \tag{8.59}$$

Substituting (8.57) into (8.59), it yields

$$\begin{aligned}\dot{V}_1 \le& -[h - (2 + 4r + 3\bar{\varsigma}(G)\bar{\varsigma}(K)\bar{\varsigma}(P))]\varepsilon^T\varepsilon + \\ &2M_\varepsilon^T M_p \tau^T\tilde{\theta} - 2M_\varepsilon^T M_p K D\tilde{b}_y + \\ &||P||^2(||\theta_\omega||^2 + \bar{d}^2 + ||\delta||^2)/r + \tilde{\theta}^T\tau\bar{P}^T\bar{P}\tau^T\tilde{\theta} + \\ &(\bar{P}\vartheta\dot{\hat{\theta}})^T\bar{P}\vartheta\dot{\hat{\theta}}/r + \tilde{b}_y^T D^T K^T\bar{P}^T P\bar{K}D\tilde{b}_y + \\ &\bar{\varsigma}(P)\bar{\varsigma}(K)\bar{\varsigma}(G)\tilde{b}_y^T\tilde{b}_y + z_1^T(B+L)z_2 - c_1 z_1^T\bar{z}_1 + \\ &c_1 z_1^T G(D+B)^{-1}\bar{z}_1 + z_1^T[(B+L)(\bar{\varepsilon}_2 + \bar{\tau}_{(1)}^T\tilde{\theta} - \\ &\hat{\bar{\gamma}} + w_{\delta 1} - w_{\theta\omega 1} + w_{d1})] - \frac{1}{\eta_\gamma}\tilde{\bar{\gamma}}^T\dot{\hat{\bar{\gamma}}} - \frac{1}{\eta_b}\tilde{b}_y^T\dot{\hat{b}}_y - \frac{1}{\eta_\theta}\tilde{\theta}^T\dot{\hat{\theta}}\end{aligned} \tag{8.60}$$

Further, one has

$$\begin{aligned}-c_1 z_1^T \bar{z}_1 &= -c_1 z_1^T [z_1 + (B+L)\tilde{b}_y] \\ &= -c_1 z_1^T z_1 - c_1 z_1^T (B+L)\tilde{b}_y \\ &\leq - c_1 z_1^T z_1 + \frac{1}{2} c_1 \bar{\varsigma}(B+L) z_1^T z_1 + \frac{1}{2} c_1 \bar{\varsigma}(B+L) \tilde{b}_y^T \tilde{b}_y\end{aligned} \tag{8.61}$$

$$\begin{aligned}& c_1 z_1^T G(D+B)^{-1}\bar{z}_1 \\ =& c_1 z_1^T G(B+D)^{-1} z_1 + c_1 z_1^T G(B+D)^{-1}(B+L)\tilde{b}_y \\ \leq& \frac{c_1 \bar{\varsigma}(G)}{\underline{\varsigma}(B+D)} z_1^T z_1 + \frac{c_1 \bar{\varsigma}(G)\bar{\varsigma}(B+L)}{2\underline{\varsigma}(B+D)} z_1^T z_1 + \frac{c_1 \bar{\varsigma}(G)\bar{\varsigma}(B+L)}{2\underline{\varsigma}(B+D)} \tilde{b}_y^T \tilde{b}_y\end{aligned} \tag{8.62}$$

$$\begin{aligned}z_1^T (B+L)\bar{\tau}_{(1)}^T \tilde{\theta} &= [\bar{z}_1 - (B+L)\tilde{b}_y]^T (B+D)\bar{\tau}_{(1)}^T \tilde{\theta} - z_1^T G \bar{\tau}_{(1)}^T \tilde{\theta} \\ &\leq \bar{z}_1^T (B+D)\bar{\tau}_{(1)}^T \tilde{\theta} + \frac{1}{2}\bar{\varsigma}(B+L)\bar{\varsigma}(B+D)\tilde{b}_y^T \tilde{b}_y + \\ &\quad \frac{1}{2}\bar{\varsigma}(B+L)\bar{\varsigma}(B+D)\tilde{\theta}^T \bar{\tau}_{(1)} \bar{\tau}_{(1)}^T \tilde{\theta} + \frac{1}{2}\bar{\varsigma}(G) z_1^T z_1 + \\ &\quad \frac{1}{2}\bar{\varsigma}(G)\tilde{\theta}^T \bar{\tau}_{(1)} \bar{\tau}_{(1)}^T \tilde{\theta}\end{aligned} \tag{8.63}$$

$$- z_1^T (B+L)\bar{\gamma} \leq \frac{1}{4}\bar{\varsigma}(B+L) z_1^T z_1 + \bar{\varsigma}(B+L)\bar{\gamma}^T \bar{\gamma} \tag{8.64}$$

$$\begin{aligned}-z_1^T (B+L)\tilde{\bar{\gamma}} &= [\bar{z}_1 - (B+L)\tilde{b}_y]^T (B+D)\tilde{\bar{\gamma}} - z_1^T G \tilde{\bar{\gamma}} \\ &\leq \bar{z}_1^T (B+D)\tilde{\bar{\gamma}} + \frac{1}{2}\bar{\varsigma}(B+L)\bar{\varsigma}(B+D)\tilde{b}_y^T \tilde{b}_y + \\ &\quad \frac{1}{2}\bar{\varsigma}(B+L)\bar{\varsigma}(B+D)\tilde{\bar{\gamma}}^T \tilde{\bar{\gamma}} + \frac{1}{2}\bar{\varsigma}(G) z_1^T z_1 + \frac{1}{2}\bar{\varsigma}(G)\tilde{\bar{\gamma}}^T \tilde{\bar{\gamma}}\end{aligned} \tag{8.65}$$

$$z_1^T (B+L)\bar{\varepsilon}_2 \leq \bar{\varsigma}(B+L) l_1 z_1^T z_1 + \frac{1}{4 l_1}\bar{\varsigma}(B+L)\bar{\varepsilon}_2^T \bar{\varepsilon}_2 \tag{8.66}$$

$$\begin{aligned}z_1^T (B+L)(w_{\delta 1} - w_{\theta\omega 1} + w_{d1}) &\leq \frac{3}{4}\bar{\varsigma}(B+L) z_1^T z_1 + \bar{\varsigma}(B+L)(||w_{\delta 1}||^2 + \\ &\quad ||w_{\theta\omega 1}||^2 + ||w_{d1}||^2) \\ &\leq \frac{3}{4}\bar{\varsigma}(B+L) z_1^T z_1 + \bar{\varsigma}(B+L) M_{\delta\omega d}\end{aligned} \tag{8.67}$$

where design parameters $c_1 > 0$ and $l_1 > 0$, $\bar{\varsigma}(\star)$ represents the maximum singular value of matrix $\star$, $M_{\delta\omega d} = ||\bar{w}_{\delta 1}||^2 + ||w_{\theta\omega 1}||^2 + ||\bar{w}_{d1}||^2$, $\bar{w}_{d1} = [\bar{d}_{1,1}, \ldots, \bar{d}_{N,1}]^T \in R^N$, $\bar{w}_{\delta 1} = [\bar{\delta}_{1,1}, \ldots, \bar{\delta}_{N,1}]^T \in R^N$.

Substituting (8.61)–(8.67) into (8.60), it yields

$$\begin{aligned}\dot{V}_1 \leq & -[h-(2+4r+3\bar{\varsigma}(G)\bar{\varsigma}(K)\bar{\varsigma}(P))]\varepsilon^T\varepsilon+2M_\varepsilon^T M_p\tau^T\tilde{\theta}- \\ & 2M_\varepsilon^T M_p K D\tilde{b}_y+\tilde{\theta}^T\tau\bar{P}^T\bar{P}\tau^T\tilde{\theta}+||P||^2(||\theta_\omega||^2+\bar{d}^2+||\bar{\delta}||^2)/r+ \\ & (\bar{P}\vartheta\dot{\hat{\theta}})^T\bar{P}\vartheta\dot{\hat{\theta}}/r+\tilde{b}_y^T D^T K^T\bar{P}^T\bar{P}KD\tilde{b}_y+\bar{\varsigma}(P)\bar{\varsigma}(K)\bar{\varsigma}(G)\tilde{b}_y^T\tilde{b}_y+ \\ & z_1^T(B+D)z_2+\frac{1}{2}\bar{\varsigma}(G)z_2^T z_2-\left[c_1-\frac{3}{2}\bar{\varsigma}(G)-\left(\frac{1}{2}c_1+l_1+1\right)\bar{\varsigma}(B+L)-\right. \\ & \left.\frac{c_1\bar{\varsigma}(G)(\bar{\varsigma}(B+L)+2)}{2\bar{\varsigma}(B+D)}\right]z_1^T z_1+\bar{z}_1^T(B+D)\bar{\tau}_{(1)}^T\tilde{\theta}+\bar{z}_1^T(B+D)\tilde{\bar{\gamma}}+ \\ & \left[\frac{1}{2}\bar{\varsigma}(B+L)\bar{\varsigma}(B+D)+\frac{1}{2}\bar{\varsigma}(G)\right]\times\tilde{\theta}^T\bar{\tau}_{(1)}\bar{\tau}_{(1)}^T\tilde{\theta}+\frac{1}{4l_1}\bar{\varsigma}(B+L)\bar{\varepsilon}_2^T\bar{\varepsilon}_2+ \\ & \left[\frac{c_1}{2}\bar{\varsigma}(B+L)\left(1+\frac{\bar{\varsigma}(G)}{\bar{\varsigma}(B+D)}\right)+\bar{\varsigma}(B+L)\bar{\varsigma}(B+D)\right]\tilde{b}_y^T\tilde{b}_y+ \\ & \bar{\varsigma}(B+L)\bar{\gamma}^T\bar{\gamma}+\frac{1}{2}[\bar{\varsigma}(B+L)\bar{\varsigma}(B+D)+\bar{\varsigma}(G)]\tilde{\bar{\gamma}}^T\tilde{\bar{\gamma}}+ \\ & \bar{\varsigma}(B+L)M_{\delta\omega d}-\frac{1}{\eta_b}\tilde{b}_y^T\dot{\hat{b}}_y-\frac{1}{\eta_\gamma}\tilde{\bar{\gamma}}^T\dot{\hat{\bar{\gamma}}}-\frac{1}{\eta_\theta}\tilde{\theta}^T\dot{\hat{\theta}}\end{aligned} \tag{8.68}$$

Here, define the following adaptive laws:

$$\dot{\hat{\theta}}=-\kappa_\theta\hat{\theta}+\eta_\theta[2\tau M_p^T M_\varepsilon+\bar{\tau}_{(1)}(B+D)\bar{z}_1] \tag{8.69}$$

$$\dot{\hat{\bar{\gamma}}}=-\kappa_\gamma\hat{\bar{\gamma}}+\eta_\gamma(B+D)\bar{z}_1 \tag{8.70}$$

$$\dot{\hat{b}}_y=-\kappa_b\hat{b}_y-2\eta_b DK^T M_p^T M_\varepsilon \tag{8.71}$$

where $\kappa_\theta > 0$, $\kappa_\gamma > 0$, $\kappa_b > 0$ are design parameters.

Substituting (8.69)–(8.71) into (8.68), we have

$$\begin{aligned}\dot{V}_1 \leq & -[h-(2+4r+3\bar{\varsigma}(G)\bar{\varsigma}(K)\bar{\varsigma}(P))]\varepsilon^T\varepsilon+ \\ & \tilde{\theta}^T\tau\bar{P}^T\bar{P}\tau^T\tilde{\theta}+\tilde{b}_y^T D^T K^T\bar{P}^T\bar{P}KD\tilde{b}_y+ \\ & \bar{\varsigma}(G)\bar{\varsigma}(K)\bar{\varsigma}(P)\tilde{b}_y^T\tilde{b}_y+\frac{\kappa_\gamma}{\eta_\gamma}\tilde{\bar{\gamma}}^T\hat{\bar{\gamma}}+\frac{\kappa_\theta}{\eta_\theta}\tilde{\theta}^T\hat{\theta}+ \\ & \frac{\kappa_b}{\eta_b}\tilde{b}_y^T\hat{b}_y+z_1^T(B+D)z_2+\frac{1}{2}\bar{\varsigma}(G)z_2^T z_2- \\ & \bar{c}_1 z_1^T z_1+[\frac{1}{2}\bar{\varsigma}(B+L)\bar{\varsigma}(B+D)+\frac{1}{2}\bar{\varsigma}(G)]\times \\ & \tilde{\theta}^T\bar{\tau}_{(1)}\bar{\tau}_{(1)}^T\tilde{\theta}+\frac{1}{4l_1}\bar{\varsigma}(B+L)\bar{\varepsilon}_2^T\bar{\varepsilon}_2+ \\ & [\bar{\varsigma}(B+L)\bar{\varsigma}(B+D)+\frac{c_1}{2}\bar{\varsigma}(B+L)(1+ \\ & \frac{\bar{\varsigma}(G)}{\bar{\varsigma}(B+D)})]\tilde{b}_y^T\tilde{b}_y+\upsilon_1+\bar{\varsigma}(B+L)\bar{\gamma}^T\bar{\gamma}+ \\ & \frac{1}{2}[\bar{\varsigma}(B+L)\bar{\varsigma}(B+D)+\bar{\varsigma}(G)]\tilde{\bar{\gamma}}^T\tilde{\bar{\gamma}}\end{aligned} \tag{8.72}$$

where $\upsilon_1 = ||P||^2(||\theta_\omega||^2 + \bar{d}^2 + ||\bar{\delta}||^2)/r + \bar{\varsigma}(B+L)M_{\delta\omega d} + (\bar{P}\vartheta\dot{\hat{\theta}})^T\bar{P}\vartheta\dot{\hat{\theta}}/r$.

Step 2: For z_2, it follows from (8.54) and (8.57) that

$$\begin{aligned}\dot{z}_2 =& \dot{w}_{m2} - \dot{\alpha}_1\\ =& w_{m3} - k_2 w_{m1} - \frac{\partial\alpha_1}{\partial y_f^T}(\bar{\xi}_2 + \bar{\vartheta}_{(2)}\hat{\theta} + \bar{w}I_m + \bar{\varepsilon}_2 +\\ &\bar{\tau}_{(1)}^T(y_f)\theta^* + w_{\delta 1} - w_{\theta\omega 1} + w_{d1}) - \frac{\partial\alpha_1}{\partial\hat{\theta}^T}\dot{\hat{\theta}} -\\ &\frac{\partial\alpha_1}{\partial\hat{\gamma}^T}\dot{\hat{\gamma}} - \sum_{r=1}^{2}\frac{\partial\alpha_1}{\partial\bar{y}_0^{T(r-1)}}\bar{y}_0^{(r)} - \frac{\partial\alpha_1}{\partial\xi^T}[A_0\xi + Ky_f +\\ &KD(y_f - \hat{y}_f)] + \frac{\partial\alpha_1}{\partial\xi^T}KG(y_f - \hat{y}_f) - \frac{\partial\alpha_1}{\partial\vartheta^T}\\ &[A_0\vartheta + \tau^T(y_f)] - \sum_{p=1}^{m+1}\frac{\partial\alpha_1}{\partial\bar{\lambda}_p^T}(-\bar{k}_p\bar{\lambda}_1 + \bar{\lambda}_{p+1})\end{aligned} \tag{8.73}$$

where $\bar{k}_p = diag\{k_{1,p}, \ldots, k_{N,p}\}$, $p = 0, \ldots, m$, $\bar{\lambda}_{p+1} = [\lambda_{1,p+1}, \ldots, \lambda_{N,p+1}]^T$.

Since $v_{m3} = z_3 + \alpha_2$, from (8.73), one has

$$\begin{aligned}\dot{z}_2 =& z_3 + \alpha_2 - \frac{\partial\alpha_1}{\partial y_f^T}(\bar{\vartheta}_{(2)}\hat{\theta} + \bar{w}I_m + \bar{\varepsilon}_2 + \bar{\tau}_{(1)}^T\theta^* +\\ &w_{\delta 1} - w_{\theta\omega 1} + w_{d1}) + \frac{\partial\alpha_1}{\partial\xi^T}KG(y_f - \hat{y}_f) + H_2\end{aligned} \tag{8.74}$$

where

$$\begin{aligned}H_2 =& -k_2 v_{m1} - \frac{\partial\alpha_1}{\partial y_f^T}\bar{\xi}_2 - \frac{\partial\alpha_1}{\partial\hat{\theta}^T}\dot{\hat{\theta}} - \frac{\partial\alpha_1}{\partial\hat{\gamma}^T}\dot{\hat{\gamma}} -\\ &\sum_{r=1}^{2}\frac{\partial\alpha_1}{\partial\bar{y}_0^{T(r-1)}}\bar{y}_0^{(r)} - \frac{\partial\alpha_1}{\partial\xi^T}[A_0\xi + Ky_f +\\ &KD(y_f - \hat{y}_f)] - \frac{\partial\alpha_1}{\partial\vartheta^T}[A_0\vartheta + \tau^T(y_f)]\\ &\sum_{p=1}^{m+1}\frac{\partial\alpha_1}{\partial\bar{\lambda}_p^T}(-\bar{k}_p\bar{\lambda}_1 + \bar{\lambda}_{p+1})\end{aligned}$$

Define α_2 as

$$\begin{aligned}\alpha_2 =& -(B+D)\bar{z}_1 - c_2 z_2 - (l_2 + \frac{3}{4})\frac{\partial\alpha_1}{\partial y_f}\times\\ &(\frac{\partial\alpha_1}{\partial y_f})^T z_2 - \frac{\partial\alpha_1}{2\partial\xi^T}(\frac{\partial\alpha_1}{\partial\xi^T})^T z_2 + \frac{\partial\alpha_1}{\partial y_f^T}\times\\ &(\bar{\vartheta}_{(2)}\hat{\theta} + \bar{w}I_m) + \frac{\partial\alpha_1}{\partial y_f^T}\bar{\tau}_{(1)}^T\hat{\theta} - H_2\end{aligned} \tag{8.75}$$

where design parameter $c_2 > 0$ satisfying

$$c_2 - \frac{1}{2}\bar{\varsigma}(G) - \frac{1}{2}\bar{\varsigma}(B+L)\bar{\varsigma}(B+D) > 0.$$

Substituting (8.75) into (8.74), we obtain

$$\begin{aligned}\dot{z}_2 =& z_3 - (B+D)\bar{z}_1 - c_2 z_2 - (l_2 + \frac{3}{4})\frac{\partial \alpha_1}{\partial y_f}(\frac{\partial \alpha_1}{\partial y_f})^T \times \\ & z_2 - \frac{\partial \alpha_1}{2\partial \xi^T}(\frac{\partial \alpha_1}{\partial \xi^T})^T z_2 - \frac{\partial \alpha_1}{\partial y_f^T}(\bar{\varepsilon}_2 + \bar{\tau}_{(1)}^T \tilde{\theta} + w_{\delta 1} - \\ & w_{\theta\omega 1} + w_{d1}) + \frac{\partial \alpha_1}{\partial \xi^T} KG(-\hat{y}_f + y_f)\end{aligned} \tag{8.76}$$

Define $V_2 = V_1 + z_2^T z_2$, then one has

$$\begin{aligned}\dot{V}_2 \le & -[h - (2 + 4r + 3\bar{\varsigma}(G)\bar{\varsigma}(K)\bar{\varsigma}(P))]\varepsilon^T\varepsilon + \\ & 2M_\varepsilon^T M_p \tau^T \tilde{\theta} - 2M_\varepsilon^T M_p K D \tilde{b}_y + \\ & \tilde{b}_y^T D^T K^T \bar{P}^T \bar{P} K D \tilde{b}_y + \tilde{\theta}^T \tau \bar{P}^T \bar{P} \tau^T \tilde{\theta} + \\ & \bar{\varsigma}(G)\bar{\varsigma}(K)\bar{\varsigma}(P)\tilde{b}_y^T \tilde{b}_y + z_1^T (B+D) z_2 + \\ & \frac{1}{2}\bar{\varsigma}(G) z_2^T z_2 - \bar{c}_1 z_1^T z_1 + \bar{z}_1^T (B+D)\bar{\tau}_{(1)}^T \tilde{\theta} + \\ & \bar{z}_1^T (B+D)\tilde{\bar{\gamma}} + \left[\frac{1}{2}\bar{\varsigma}(B+L)\bar{\varsigma}(B+D) + \right. \\ & \frac{1}{4l_1}\bar{\varsigma}(B+L)\bar{\varepsilon}_2^T\bar{\varepsilon}_2 + \frac{1}{2}\bar{\varsigma}(G)]\tilde{\theta}^T \bar{\tau}_{(1)}\bar{\tau}_{(1)}^T \tilde{\theta} + \\ & \left[\frac{c_1}{2}\bar{\varsigma}(B+L)\left(1 + \frac{\bar{\varsigma}(G)}{\bar{\varsigma}(B+D)}\right) + \bar{\varsigma}(B+L) \times \right. \\ & \bar{\varsigma}(B+D)]\tilde{b}_y^T \tilde{b}_y + \bar{\varsigma}(B+L)\tilde{\gamma}^T \tilde{\gamma} + \frac{1}{2}[\bar{\varsigma}(G) + \\ & \left.\bar{\varsigma}(B+L)\bar{\varsigma}(B+D)\right]\tilde{\bar{\gamma}}^T \tilde{\bar{\gamma}} + \upsilon_1 + z_2^T [z_3 - \\ & (B+D)\bar{z}_1 - c_2 z_2 - \left(l_2 + \frac{3}{4}\right)\frac{\partial \alpha_1}{\partial y_f}(\frac{\partial \alpha_1}{\partial y_f})^T z_2 - \\ & \frac{\partial \alpha_1}{2\partial \xi^T}\left(\frac{\partial \alpha_1}{\partial \xi^T}\right)^T z_2 - \frac{\partial \alpha_1}{\partial y_f^T}(\bar{\varepsilon}_2 + \bar{\tau}_{(1)}^T \tilde{\theta} + w_{\delta 1} - \\ & \left. w_{\theta\omega 1} + w_{d1}) + \frac{\partial \alpha_1}{\partial \xi^T} KG(y_f - \hat{y}_f)\right] - \frac{1}{\eta_\theta}\tilde{\theta}^T \dot{\hat{\theta}} - \\ & \frac{1}{\eta_\gamma}\tilde{\bar{\gamma}}^T \dot{\hat{\bar{\gamma}}} - \frac{1}{\eta_b}\tilde{b}_y^T \dot{\hat{b}}_y\end{aligned} \tag{8.77}$$

Based on Young's inequality, it follows

$$
\begin{aligned}
-z_2^T(B+D)\bar{z}_1 &= -z_2^T(B+D)[z_1+(B+L)\tilde{b}_y] \\
&= -z_2^T(B+D)z_1 - z_2^T(B+D)(B+L)\tilde{b}_y \\
&\leq -z_2^T(B+D)z_1 + \frac{1}{2}\varsigma(B+D)\varsigma(B+L)\times \\
&\quad z_2^T z_2 + \frac{1}{2}\varsigma(B+D)\varsigma(B+L)\tilde{b}_y^T\tilde{b}_y
\end{aligned}
\tag{8.78}
$$

$$
-z_2^T\frac{\partial\alpha_1}{\partial y_f^T}\bar{\varepsilon}_2 \leq l_2 z_2^T\frac{\partial\alpha_1}{\partial y_f^T}(\frac{\partial\alpha_1}{\partial y_f^T})^2 z_2 + \frac{1}{4l_2}\bar{\varepsilon}_2^T\bar{\varepsilon}_2
\tag{8.79}
$$

$$
-z_2^T\frac{\partial\alpha_1}{\partial y_f^T}(w_{\delta 1} - w_{\theta\omega 1} + w_{d1}) \leq \frac{3}{4}z_2^T\frac{\partial\alpha_1}{\partial y_f^T}\left(\frac{\partial\alpha_1}{\partial y_f^T}\right)z_2 + M_{\delta\omega d}
\tag{8.80}
$$

$$
\begin{aligned}
z_2^T\frac{\partial\alpha_1}{\partial\xi^T}KG(y_f-\hat{y}_f) \leq& \frac{1}{2}z_2^T\frac{\partial\alpha_1}{\partial\xi^T}\left(\frac{\partial\alpha_1}{\partial\xi^T}\right)^T z_2 + \\
&\bar{\varsigma}^2(K)\bar{\varsigma}^2(G)\varepsilon^T\varepsilon + \bar{\varsigma}^2(K)\bar{\varsigma}^2(G)\tilde{b}_y^T\tilde{b}_y
\end{aligned}
\tag{8.81}
$$

Substituting (8.78)–(8.81) into (8.77), we obtain

$$
\begin{aligned}
\dot{V}_2 \leq& -[h-(2+4r2+3\bar{\varsigma}(G)\bar{\varsigma}(K)\bar{\varsigma}(P)+\bar{\varsigma}^2(K)\bar{\varsigma}^2(G))]\varepsilon^T\varepsilon + 2M_\varepsilon^T M_p\tau^T\tilde{\theta} - \\
&2M_\varepsilon^T M_p K D\tilde{b}_y + \tilde{b}_y^T D^T K^T \bar{P}^T\bar{P}KD\tilde{b}_y + \tilde{\theta}^T\tau\bar{P}^T\bar{P}\tau^T\tilde{\theta} + \bar{\varsigma}(G)\bar{\varsigma}(K)\bar{\varsigma}(P)\tilde{b}_y^T\tilde{b}_y - \\
&\bar{c}_1 z_1^T z_1 + \bar{z}_1^T(B+D)\bar{\tau}_{(1)}^T\tilde{\theta} + \bar{z}_1^T(B+D)\tilde{\bar{\gamma}} + \frac{1}{4l_1}\bar{\varsigma}(B+L)\bar{\varepsilon}_2^T\bar{\varepsilon}_2 + \\
&\left[\frac{1}{2}\bar{\varsigma}(B+L)\bar{\varsigma}(B+D) + \frac{1}{2}\bar{\varsigma}(G)\right]\tilde{\theta}^T\bar{\tau}_{(1)}\bar{\tau}_{(1)}^T\tilde{\theta} + \\
&\left[\frac{c_1}{2}\bar{\varsigma}(B+L)\left(1+\frac{\bar{\varsigma}(G)}{\bar{\varsigma}(B+D)}\right) + \bar{\varsigma}(B+L)\bar{\varsigma}(B+D)\right]\tilde{b}_y^T\tilde{b}_y + \\
&\bar{\varsigma}(B+L)\tilde{\gamma}^T\tilde{\gamma} + \frac{1}{2}[\bar{\varsigma}(B+L)\bar{\varsigma}(B+D)+\bar{\varsigma}(G)]\times\tilde{\bar{\gamma}}^T\tilde{\bar{\gamma}} - \\
&\left[c_2 - \frac{1}{2}\bar{\varsigma}(G) - \frac{1}{2}\bar{\varsigma}(B+L)\bar{\varsigma}(B+D)\right]\times z_2^T z_2 + z_2^T z_3 - z_2^T\frac{\partial\alpha_1}{\partial y_f^T}\bar{\tau}_{(1)}^{\prime}\theta + \\
&\left[\frac{1}{2}\varsigma(B+D)\varsigma(B+L) + \bar{\varsigma}^2(G)\bar{\varsigma}^2(K)\right]\tilde{b}_y^T\tilde{b}_y + \frac{1}{4l_2}\bar{\varepsilon}_2^T\bar{\varepsilon}_2 + \\
&\upsilon_2 - \frac{1}{\eta_\theta}\tilde{\theta}^T\dot{\hat{\theta}} - \frac{1}{\eta_\gamma}\tilde{\bar{\gamma}}^T\dot{\hat{\bar{\gamma}}} - \frac{1}{\eta_b}\tilde{b}_y^T\dot{\hat{b}}_y
\end{aligned}
\tag{8.82}
$$

where $\upsilon_2 = \upsilon_1 + M_{\delta\omega d}$. Define the adaptive laws

$$
\dot{\hat{\theta}} = \eta_\theta[2\tau M_p^T M_\varepsilon + \bar{\tau}_{(1)}(B+D)\bar{z}_1 - \bar{\tau}_{(1)}\frac{\partial\alpha_1}{\partial y_f^T}z_2] - \kappa_\theta\hat{\theta}
\tag{8.83}
$$

$$
\dot{\hat{\bar{\gamma}}} = \eta_\gamma(B+D)\bar{z}_1 - \kappa_\gamma\hat{\bar{\gamma}}
\tag{8.84}
$$

$$\dot{\hat{b}}_y = -2\eta_b D K^T M_p^T M_\varepsilon - \kappa_b \hat{b}_y \tag{8.85}$$

where $\kappa_\theta > 0$, $\kappa_\gamma > 0$, $\kappa_b > 0$ are design parameters.

Substituting (8.83)–(8.85) into (8.82), it yields

$$\begin{aligned}
\dot{V}_2 \leq & -[h - (2 + 4r + 3\bar{\varsigma}(G)\bar{\varsigma}(K)\bar{\varsigma}(P) + \bar{\varsigma}^2(K)\bar{\varsigma}^2(G))]\varepsilon^T\varepsilon + \\
& \left[\frac{1}{4l_1}\bar{\varsigma}(B+L) + \frac{1}{4l_2}\right]\bar{\varepsilon}_2^T\bar{\varepsilon}_2 + \tilde{\theta}^T\tau\bar{P}^T\bar{P}\tau^T\tilde{\theta} + \tilde{b}_y^T D^T K^T \bar{P}^T \bar{P} K D \tilde{b}_y + \\
& \bar{\varsigma}(G)\bar{\varsigma}(K)\bar{\varsigma}(P)\tilde{b}_y^T\tilde{b}_y - \bar{c}_1 z_1^T z_1 - \bar{c}_2 z_2^T z_2 + \\
& \left[\frac{1}{2}\bar{\varsigma}(B+L)\bar{\varsigma}(B+D) + \frac{1}{2}\bar{\varsigma}(G)\right]\tilde{\theta}^T\bar{\tau}_{(1)}\bar{\tau}_{(1)}^T\tilde{\theta} + \\
& z_2^T z_3 + \bar{\varsigma}(B+L)\tilde{\gamma}^T\tilde{\gamma} + \frac{1}{2}[\bar{\varsigma}(B+L)\bar{\varsigma}(B+D) + \bar{\varsigma}(G)]\tilde{\bar{\gamma}}^T\tilde{\bar{\gamma}} + \\
& \left[\frac{c_1}{2}\bar{\varsigma}(B+L)(1 + \frac{\bar{\varsigma}(G)}{\bar{\varsigma}(B+D)}) + \frac{3}{2}\bar{\varsigma}(B+L)\bar{\varsigma}(B+D) + \bar{\varsigma}^2(K)\bar{\varsigma}^2(G)\right]\tilde{b}_y^T\tilde{b}_y + \\
& \frac{\kappa_\theta}{\eta_\theta}\tilde{\theta}^T\hat{\theta} + \frac{\kappa_\gamma}{\eta_\gamma}\tilde{\bar{\gamma}}^T\hat{\bar{\gamma}} + \frac{\kappa_b}{\eta_b}\tilde{b}_y^T\hat{b}_y + \upsilon_2
\end{aligned} \tag{8.86}$$

Step q ($q = 3, 4, \ldots, \rho$): From (8.54) and (8.82), one has

$$\begin{aligned}
\dot{z}_q = & w_{m(q+1)} - k_q w_{m1} - \frac{\partial\alpha_{q-1}}{\partial y_f^T}(\bar{\xi}_2 + \bar{\vartheta}_{(2)}\hat{\theta} + \bar{w}I_m + \\
& \bar{\varepsilon}_2 + \bar{\tau}_{(1)}^T(y_f)\theta^* + w_{\delta 1} - w_{\theta\omega 1} + w_{d1}) - \frac{\partial\alpha_{q-1}}{\partial\hat{\theta}^T}\dot{\hat{\theta}} - \frac{\partial\alpha_{q-1}}{\partial\hat{\gamma}^T}\dot{\hat{\gamma}} - \\
& \sum_{r=1}^{q}\frac{\partial\alpha_{q-1}}{\partial\bar{y}_0^{T(r-1)}}\bar{y}_0^{(r)} - \frac{\partial\alpha_{q-1}}{\partial\xi^T}[A_0\xi + K y_f + K D(y_f - \hat{y}_f)] + \\
& \frac{\partial\alpha_{q-1}}{\partial\xi^T}K G(y_f - \hat{y}_f) - \frac{\partial\alpha_{q-1}}{\partial\vartheta^T}[A_0\vartheta + \tau^T] - \sum_{p=1}^{m+q-1}\frac{\partial\alpha_{q-1}}{\partial\bar{\lambda}_p^T}(-\bar{k}_p\bar{\lambda}_1 + \bar{\lambda}_{p+1})
\end{aligned} \tag{8.87}$$

Because $w_{m(q+1)} = z_{q+1} + \alpha_q$, from (8.87), we have

$$\begin{aligned}
\dot{z}_q = & z_{q+1} + \alpha_q + \frac{\partial\alpha_{q-1}}{\partial\xi^T}K G(y_f - \hat{y}_f) + H_q - \\
& \frac{\partial\alpha_{q-1}}{\partial y_f^T}(\bar{\vartheta}_{(2)}\hat{\theta} + \bar{w}I_m + \bar{\varepsilon}_2 + \bar{\tau}_{(1)}^T(y_f)\theta^* + w_{\delta 1} - w_{\theta\omega 1} + w_{d1})
\end{aligned} \tag{8.88}$$

where

$$H_q = -k_q w_{m1} - \frac{\partial \alpha_{q-1}}{\partial y_f^T}\bar{\xi}_2 - \frac{\partial \alpha_{q-1}}{\partial \hat{\theta}^T}\dot{\hat{\theta}} - \frac{\partial \alpha_{q-1}}{\partial \hat{\gamma}^T}\dot{\hat{\gamma}} -$$
$$\sum_{r=1}^{q} \frac{\partial \alpha_{q-1}}{\partial \bar{y}_0^{T(r-1)}}\bar{y}_0^{(r)} - \frac{\partial \alpha_{q-1}}{\partial \xi^T}[A_0\xi + Ky_f + KD(y_f - \hat{y}_f)] -$$
$$\frac{\partial \alpha_{q-1}}{\partial \vartheta^T}[A_0\vartheta + \tau^T(y_f)] - \sum_{p=1}^{m+q-1} \frac{\partial \alpha_{q-1}}{\partial \bar{\lambda}_p^T}(-\bar{k}_p\bar{\lambda}_1 + \bar{\lambda}_{p+1})$$

Now, α_q is defined as

$$\begin{aligned} \alpha_q = & -z_{q-1} - c_q z_q - (l_q + \frac{3}{4})\frac{\partial \alpha_{q-1}}{\partial y_f}(\frac{\partial \alpha_{q-1}}{\partial y_f})^T z_q - \\ & \frac{\partial \alpha_{q-1}}{2\partial \xi^T}(\frac{\partial \alpha_{q-1}}{\partial \xi^T})^T z_q + \frac{\partial \alpha_{q-1}}{\partial y_f^T}(\bar{\vartheta}_{(2)}\hat{\theta} + \bar{w}I_m) + \frac{\partial \alpha_{q-1}}{\partial y_f^T}\bar{\tau}_{(1)}^T\hat{\theta} - H_q \end{aligned} \tag{8.89}$$

From (8.51) and (8.89), we have

$$\alpha_\rho = u + w_{m(\rho+1)}, z_{\rho+1} = [0, \ldots, 0]^T \tag{8.90}$$

Substituting (8.89) into (8.88), one has

$$\begin{aligned} \dot{z}_q = & -z_{q-1} - c_q z_q + z_{q+1} - (l_q + \frac{3}{4})\frac{\partial \alpha_{q-1}}{\partial y_f} \\ & (\frac{\partial \alpha_{q-1}}{\partial y_f})^T z_q - \frac{\partial \alpha_{q-1}}{2\partial \xi^T}(\frac{\partial \alpha_{q-1}}{\partial \xi^T})^T z_q - \\ & \frac{\partial \alpha_{q-1}}{\partial y_f^T}(\bar{\varepsilon}_2 + \bar{\tau}_{(1)}^T\tilde{\theta} + w_{\delta 1} - w_{\theta\omega 1} + w_{d1}) + \frac{\partial \alpha_{q-1}}{\partial \xi^T}KG(y_f - \hat{y}_f) \end{aligned} \tag{8.91}$$

Define

$$V_q = V_{q-1} + \frac{1}{2}z_q^T z_q \tag{8.92}$$

From (8.91), $\dot{V}_q$ is developed as

$$
\begin{aligned}
\dot{V}_q \leq & -[h-(2+4r+3\bar{\varsigma}(G)\bar{\varsigma}(K)\bar{\varsigma}(P)+\\
& (q-2)\bar{\varsigma}^2(K)\bar{\varsigma}^2(G))]\varepsilon^T\varepsilon+[\frac{\bar{\varsigma}(B+L)}{4l_1}+\frac{1}{4l_2}+\\
& \cdots+\frac{1}{4l_{q-1}}]\bar{\varepsilon}_2^T\bar{\varepsilon}_2-\bar{c}_1z_1^Tz_1-\bar{c}_2z_2^Tz_2-\cdots-\\
& c_{q-1}z_{q-1}^Tz_{q-1}+\frac{1}{\eta_\theta}\tilde{\theta}^T[\eta_\theta(2\tau M_p^TM_\varepsilon+\\
& \bar{\tau}_{(1)}(B+D)\bar{z}_1-\cdots-\bar{\tau}_{(1)}\frac{\partial\alpha_{q-2}}{\partial y_f^T}z_{q-1})-\dot{\hat{\theta}}]+\\
& \frac{1}{\eta_\gamma}\tilde{\gamma}^T[\eta_\gamma(B+D)\bar{z}_1-\dot{\hat{\gamma}}]+\tilde{\theta}^T\tau\bar{P}^T\bar{P}\tau^T\tilde{\theta}+\\
& \frac{1}{\eta_b}\tilde{b}_y^T[-2\eta_bDK^TM_p^TM_\varepsilon-\dot{\hat{b}}_y]+[\frac{1}{2}\bar{\varsigma}(B+L)\\
& \bar{\varsigma}(B+D)+\frac{1}{2}\bar{\varsigma}(G)]\tilde{\theta}^T\bar{\tau}_{(1)}\bar{\tau}_{(1)}^T\tilde{\theta}+\tilde{b}_y^TD^TK^T\times\\
& \bar{P}^T\bar{P}KD\tilde{b}_y+\bar{\varsigma}(G)\bar{\varsigma}(K)\bar{\varsigma}(P)\tilde{b}_y^T\tilde{b}_y+\\
& [\frac{c_1}{2}\bar{\varsigma}(B+L)(1+\frac{\bar{\varsigma}(G)}{\bar{\varsigma}(B+D)})+\frac{3}{2}\bar{\varsigma}(B+L)\\
& \bar{\varsigma}(B+D)+(q-2)\bar{\varsigma}^2(K)\bar{\varsigma}^2(G)]\tilde{b}_y^T\tilde{b}_y+\\
& \bar{\varsigma}(B+L)\bar{\gamma}^T\bar{\gamma}+\frac{1}{2}[\bar{\varsigma}(B+L)\bar{\varsigma}(B+D)+\\
& \bar{\varsigma}(G)]\bar{\tilde{\gamma}}^T\bar{\tilde{\gamma}}+z_{q-1}^Tz_q+\upsilon_{q-1}+z_q^T[-z_{q-1}-\\
& c_qz_q+z_{q+1}-(l_q+\frac{3}{4})\frac{\partial\alpha_{q-1}}{\partial y_f^T}(\frac{\partial\alpha_{q-1}}{\partial y_f^T})^Tz_q-\\
& \frac{\partial\alpha_{q-1}}{2\partial\xi^T}(\frac{\partial\alpha_{q-1}}{\partial\xi^T})^Tz_q-\frac{\partial\alpha_{q-1}}{\partial y_f^T}(\bar{\varepsilon}_2+\bar{\tau}_{(1)}^T\tilde{\theta}+\\
& w_{\delta 1}-w_{\theta\omega 1}+w_{d1})+\frac{\partial\alpha_{q-1}}{\partial\xi^T}KG(y_f-\hat{y}_f)]
\end{aligned}
\tag{8.93}
$$

From Young's inequality, it yields

$$
z_q^T\frac{\partial\alpha_{q-1}}{\partial y_f^T}\bar{\varepsilon}_2\leq l_qz_q^T\frac{\partial\alpha_{q-1}}{\partial y_f^T}(\frac{\partial\alpha_{q-1}}{\partial y_f^T})^Tz_q+\frac{1}{4l_q}\bar{\varepsilon}_2^T\bar{\varepsilon}_2 \tag{8.94}
$$

$$
-z_q^T\frac{\partial\alpha_{q-1}}{\partial y_f^T}(w_{\delta 1}-w_{\theta\omega 1}+w_{d1})\leq\frac{3}{4}z_q^T\frac{\partial\alpha_{q-1}}{\partial y_f^T}(\frac{\partial\alpha_{q-1}}{\partial y_f^T})^Tz_q+M_{\delta\omega d} \tag{8.95}
$$

$$z_q^T \frac{\partial \alpha_{q-1}}{\partial \xi^T} KG(y_f - \hat{y}_f) \leq \frac{1}{2} z_q^T \frac{\partial \alpha_{q-1}}{\partial \xi^T} (\frac{\partial \alpha_{q-1}}{\partial \xi^T})^T z_q + \\ \bar{\varsigma}^2(G)\bar{\varsigma}^2(K)\varepsilon^T \varepsilon + \bar{\varsigma}^2(G)\bar{\varsigma}^2(K)\tilde{b}_y^T \tilde{b}_y \tag{8.96}$$

From (8.93)–(8.96), if follows

$$\begin{aligned} \dot{V}_q \leq & -[h - (2 + 4r + 3\bar{\varsigma}(G)\bar{\varsigma}(K)\bar{\varsigma}(P) + \\ & (q-1)\bar{\varsigma}^2(K)\bar{\varsigma}^2(G))]\varepsilon^T \varepsilon + [\frac{\bar{\varsigma}(B+L)}{4l_1} + \frac{1}{4l_2} + \\ & \cdots + \frac{1}{4l_q}]\bar{\varepsilon}_2^T \bar{\varepsilon}_2 - \bar{c}_1 z_1^T z_1 - \cdots - c_q z_q^T z_q + \\ & \frac{1}{\eta_\theta}\tilde{\theta}^T [\eta_\theta (2\tau M_p^T M_\varepsilon + \bar{\tau}_{(1)}(B+D)\bar{z}_1 - \cdots - \\ & \bar{\tau}_{(1)} \frac{\partial \alpha_{q-1}}{\partial y_f^T} z_q) - \dot{\hat{\theta}}] + \frac{1}{\eta_\gamma}\tilde{\gamma}^T [\eta_\gamma (B+D)\bar{z}_1 - \dot{\hat{\gamma}}] + \\ & \frac{1}{\eta_b}\tilde{b}_y^T [-2\eta_b D K^T M_p^T M_\varepsilon - \dot{\hat{b}}_y] + \tilde{\theta}^T \tau \bar{P}^T \bar{P} \tau^T \tilde{\theta} + \\ & [\frac{1}{2}\bar{\varsigma}(B+L)\bar{\varsigma}(B+D) + \frac{1}{2}\bar{\varsigma}(G)]\tilde{\theta}^T \bar{\tau}_{(1)} \bar{\tau}_{(1)}^T \tilde{\theta} + \\ & \tilde{b}_y^T D^T K^T \bar{P}^T \bar{P} K D \tilde{b}_y + \bar{\varsigma}(P)\bar{\varsigma}(K)\bar{\varsigma}(G)\tilde{b}_y^T \tilde{b}_y + \\ & [\frac{c_1}{2}\bar{\varsigma}(B+L)(1 + \frac{\bar{\varsigma}(G)}{\bar{\varsigma}(B+D)}) + \frac{3}{2}\bar{\varsigma}(B+L) \\ & \bar{\varsigma}(B+D) + (q-1)\bar{\varsigma}^2(K)\bar{\varsigma}^2(G)]\tilde{b}_y^T \tilde{b}_y + \\ & \bar{\varsigma}(B+L)\tilde{\gamma}^T \tilde{\gamma} + \frac{1}{2}[\bar{\varsigma}(B+L)\bar{\varsigma}(B+D) + \\ & \bar{\varsigma}(G)]\tilde{\bar{\gamma}}^T \tilde{\bar{\gamma}} + z_q^T z_{q+1} + \upsilon_q \end{aligned} \tag{8.97}$$

where $\upsilon_q = \upsilon_{q-1} + M_{\delta\omega d}$.

Now, design the following control and adaptive laws

$$u = \alpha_\rho - w_{m(\rho+1)} \tag{8.98}$$

$$\dot{\hat{\theta}} = \eta_\theta (2\tau M_p^T M_\varepsilon + \iota_\rho) - \kappa_\theta \hat{\theta} \tag{8.99}$$

$$\dot{\hat{\bar{\gamma}}} = \eta_\gamma (B+D)\bar{z}_1 - \kappa_\gamma \hat{\bar{\gamma}} \tag{8.100}$$

$$\dot{\hat{b}}_y = -2\eta_b D K^T M_p^T M_\varepsilon - \kappa_b \hat{b}_y \tag{8.101}$$

where $\kappa_\theta > 0$, $\kappa_\gamma > 0$, $\kappa_b > 0$ are design parameters.

From (8.99), τ_ρ is chosen as

$$\tau_1 = \bar{\tau}_{(1)}(B + D)\bar{z}_1 \tag{8.102}$$

$$\tau_2 = \tau_1 - \bar{\tau}_{(1)}\frac{\partial \alpha_1}{\partial y_f^T} z_2 \tag{8.103}$$

$$\tau_q = \tau_{q-1} - \bar{\tau}_{(1)}\frac{\partial \alpha_{q-1}}{\partial y_f^T} z_q \tag{8.104}$$

Substituting (8.98)–(8.101) into (8.97), one has

$$\begin{aligned}
\dot{V}_q \leq & -[h - (2 + 4r + 3\bar{\varsigma}(G)\bar{\varsigma}(K)\bar{\varsigma}(P) + \\
& (q-1)\bar{\varsigma}^2(G)\bar{\varsigma}^2(K))]\varepsilon^T\varepsilon + [\frac{\bar{\varsigma}(B+L)}{4l_1} + \frac{1}{4l_2} + \\
& \cdots + \frac{1}{4l_q}]\bar{\varepsilon}_2^T\bar{\varepsilon}_2 - \bar{c}_1 z_1^T z_1 - \cdots - \bar{c}_q z_q^T z_q + \\
& \tilde{\theta}^T[\tau\bar{P}^T\bar{P}\tau^T + (\frac{1}{2}\bar{\varsigma}(B+L)\bar{\varsigma}(B+D) + \\
& \frac{1}{2}\bar{\varsigma}(G))\bar{\tau}_{(1)}\bar{\tau}_{(1)}^T]\tilde{\theta} + \tilde{b}_y^T D^T K^T \bar{P}^T \bar{P} K D \tilde{b}_y + \\
& \bar{\varsigma}(G)\bar{\varsigma}(K)\bar{\varsigma}(P)\tilde{b}_y^T\tilde{b}_y + [\frac{c_1}{2}\bar{\varsigma}(B+L)(1 + \\
& \frac{\bar{\varsigma}(G)}{\underline{\varsigma}(B+D)}) + \frac{3}{2}\bar{\varsigma}(B+L)\bar{\varsigma}(B+D) + (q-1) \\
& \bar{\varsigma}^2(G)\bar{\varsigma}^2(K)]\tilde{b}_y^T\tilde{b}_y + \frac{1}{2}[\bar{\varsigma}(B+L)\bar{\varsigma}(B+D) + \\
& \bar{\varsigma}(G)]\tilde{\bar{\gamma}}^T\tilde{\bar{\gamma}} + \bar{\varsigma}(B+L)\tilde{\gamma}^T\tilde{\gamma} + z_q^T z_{q+1} + \upsilon_q + \\
& \frac{\kappa_b}{\eta_b}\tilde{b}_y^T\hat{b}_y + \frac{\kappa_\gamma}{\eta_\gamma}\tilde{\bar{\gamma}}^T\hat{\bar{\gamma}} + \frac{\kappa_\theta}{\eta_\theta}\tilde{\theta}^T\hat{\theta}
\end{aligned} \tag{8.105}$$

8.3.3 Stability Analysis

Define

$$V = V_\varepsilon + \frac{1}{2}\sum_{q=1}^{\rho} z_q^T z_q + \frac{1}{2\eta_\theta}\tilde{\theta}^T\tilde{\theta} + \frac{1}{2\eta_\gamma}\tilde{\bar{\gamma}}^T\tilde{\bar{\gamma}} + \frac{1}{\eta_b}\tilde{b}_y^T\tilde{b}_y \tag{8.106}$$

From (8.98), (8.99)–(8.101), $\dot{V}$ can be developed as

$$
\begin{aligned}
\dot{V} \leq & -[h-(2+4r+3\bar{\varsigma}(G)\bar{\varsigma}(K)\bar{\varsigma}(P)+ \\
& (\rho-1)\bar{\varsigma}^2(G)\bar{\varsigma}^2(K))]\varepsilon^T\varepsilon+[\frac{\bar{\varsigma}(B+L)}{4l_1}+ \\
& \frac{1}{4l_2}+\cdots+\frac{1}{4l_\rho}]\bar{\varepsilon}_2^T\bar{\varepsilon}_2-\bar{c}_1 z_1^T z_1-\cdots- \\
& c_\rho z_\rho^T z_\rho+\tilde{\theta}^T[\tau\bar{P}^T\bar{P}\tau^T+(\frac{1}{2}\bar{\varsigma}(B+L) \\
& \bar{\varsigma}(B+D)+\frac{1}{2}\bar{\varsigma}(G))\bar{\tau}_{(1)}\bar{\tau}_{(1)}^T]\tilde{\theta}+\tilde{b}_y^T D^T K^T\times \\
& \bar{P}^T\bar{P}KD\tilde{b}_y+\bar{\varsigma}(G)\bar{\varsigma}(K)\bar{\varsigma}(P)\tilde{b}_y^T\tilde{b}_y+[\frac{c_1}{2}\times \\
& \bar{\varsigma}(B+L)(1+\frac{\bar{\varsigma}(G)}{\bar{\varsigma}(B+D)})+\frac{3}{2}\bar{\varsigma}(B+L) \\
& \bar{\varsigma}(B+D)+(\rho-1)\bar{\varsigma}^2(G)\bar{\varsigma}^2(K)]\tilde{b}_y^T\tilde{b}_y+ \\
& \frac{1}{2}[\bar{\varsigma}(B+L)\bar{\varsigma}(B+D)+\bar{\varsigma}(G)]\tilde{\bar{\gamma}}^T\tilde{\bar{\gamma}}+\bar{\varsigma}(B+L) \\
& \frac{\kappa_b}{\eta_b}\tilde{b}_y^T\hat{b}_y+\frac{\kappa_\gamma}{\eta_\gamma}\tilde{\bar{\gamma}}^T\hat{\bar{\gamma}}+\frac{\kappa_\theta}{\eta_\theta}\tilde{\theta}^T\hat{\theta}+\bar{\gamma}^T\bar{\gamma}+\upsilon_\rho+
\end{aligned}
\tag{8.107}
$$

Let $h=2+4r+3\bar{\varsigma}(G)\bar{\varsigma}(K)\bar{\varsigma}(P)+(\rho-1)\bar{\varsigma}^2(G)\bar{\varsigma}^2(K)+\frac{\bar{\varsigma}(B+L)}{2l_1}+\sum_{q=2}^{\rho}\frac{1}{2l_q}$.

Because

$$
\begin{aligned}
\frac{\kappa_\theta}{\eta_\theta}\tilde{\theta}^T\hat{\theta} &= \frac{\kappa_\theta}{\eta_\theta}\tilde{\theta}^T(\theta^*-\tilde{\theta}) \\
&= \frac{\kappa_\theta}{\eta_\theta}\tilde{\theta}^T\theta^*-\frac{\kappa_\theta}{\eta_\theta}\tilde{\theta}^T\tilde{\theta} \\
&\leq -\frac{\kappa_\theta}{2\eta_\theta}\tilde{\theta}^T\tilde{\theta}+\frac{\kappa_\theta}{2\eta_\theta}||\theta^*||^2
\end{aligned}
\tag{8.108}
$$

$$
\frac{\kappa_\gamma}{\eta_\gamma}\tilde{\bar{\gamma}}^T\hat{\bar{\gamma}} \leq -\frac{\kappa_\gamma}{2\eta_\gamma}\tilde{\bar{\gamma}}^T\tilde{\bar{\gamma}}+\frac{\kappa_\gamma}{2\eta_\gamma}\bar{\gamma}^T\bar{\gamma}
\tag{8.109}
$$

$$
\frac{\kappa_b}{\eta_b}\tilde{b}_y^T\hat{b}_y \leq -\frac{\kappa_b}{2\eta_b}\tilde{b}_y^T\tilde{b}_y+\frac{\kappa_b}{2\eta_b}b_0^T b_0
\tag{8.110}
$$

where $b_0=[b_{1,0},\ldots,b_{N,0}]^T \in R^N$.

Substituting (8.108)–(8.110) into (8.107), we can further obtain

$$\begin{aligned}\dot{V} \leq &- l_\varepsilon \varepsilon^T \varepsilon - \bar{c}_1 z_1^T z_1 - \cdots - c_\rho z_\rho^T z_\rho + \\ &\tilde{\theta}^T [\tau \bar{P}^T \bar{P} \tau^T + (\frac{1}{2} \bar{\varsigma}(B+L) \varsigma(B+D) + \\ &\frac{1}{2} \bar{\varsigma}(G)) \bar{\tau}_{(1)} \bar{\tau}_{(1)}^T] \tilde{\theta} + \tilde{b}_y^T D^T K^T \bar{P}^T \bar{P} K D \tilde{b}_y + \\ &\bar{\varsigma}(G) \bar{\varsigma}(K) \bar{\varsigma}(P) \tilde{b}_y^T \tilde{b}_y + [\frac{c_1}{2} \bar{\varsigma}(B+L)(1 + \\ &\frac{\bar{\varsigma}(G)}{\varsigma(B+D)}) + \frac{3}{2} \bar{\varsigma}(B+L) \varsigma(B+D) + (\rho - 1) \\ &\bar{\varsigma}^2(G) \bar{\varsigma}^2(K)] \tilde{b}_y^T \tilde{b}_y + \frac{1}{2} [\bar{\varsigma}(B+L) \varsigma(B+D) + \\ &\bar{\varsigma}(G)] \tilde{\bar{\gamma}}^T \tilde{\bar{\gamma}} + \upsilon_0 - \frac{\kappa_b}{2\eta_b} \tilde{b}_y^T \tilde{b}_y - \frac{\kappa_\gamma}{2\eta_\gamma} \tilde{\bar{\gamma}}^T \tilde{\bar{\gamma}} - \frac{\kappa_\theta}{2\eta_\theta} \tilde{\theta}^T \tilde{\theta}\end{aligned} \tag{8.111}$$

where $l_\varepsilon = \frac{\bar{\varsigma}(B+L)}{2l_1} + \sum_{q=2}^{\rho} \frac{1}{2l_q}$, $\upsilon_0 = \upsilon_\rho + \bar{\varsigma}(B+L) \bar{\gamma}^T \bar{\gamma} + \frac{\kappa_\theta}{\eta_\theta} ||\theta^*||^2 + \frac{\kappa_\gamma}{2\eta_\gamma} \bar{\gamma}^T \bar{\gamma} + \frac{\kappa_b}{2\eta_b} b_0^T b_0$.

Let

$$\kappa_\theta I_N = \eta_\theta [2\tau \bar{P}^T \bar{P} \tau^T + \bar{\varsigma}(B+L) \varsigma(B+D) + \bar{\varsigma}(G)) \bar{\tau}_{(1)} \bar{\tau}_{(1)}^T] + I_N$$

$$\kappa_\gamma = \eta_\gamma [\bar{\varsigma}(B+L) \varsigma(B+D) + \bar{\varsigma}(G)] + 1$$

$$\begin{aligned}\kappa_b I_N = &\eta_b \{2 D^T K^T \bar{P}^T \bar{P} K D + 2 \bar{\varsigma}(P) \bar{\varsigma}(K) \bar{\varsigma}(G) + c_1 \bar{\varsigma}(B+L)(1 + \frac{\bar{\varsigma}(G)}{\varsigma(B+D)}) + \\ &3 \bar{\varsigma}(B+L) \varsigma(B+D) + 2(\rho - 1) \bar{\varsigma}^2(G) \bar{\varsigma}^2(K)\} + I_N,\end{aligned}$$

then it follows

$$\begin{aligned}\dot{V} \leq &- l_\varepsilon \varepsilon^T \varepsilon - \bar{c}_1 z_1^T z_1 - \cdots - \bar{c}_\rho z_\rho^T z_\rho - \\ &\frac{1}{2\eta_b} \tilde{b}_y^T \tilde{b}_y + \upsilon_0 - \frac{1}{2\eta_\gamma} \tilde{\bar{\gamma}}^T \tilde{\bar{\gamma}} - \frac{1}{2\eta_\theta} \tilde{\theta}^T \tilde{\theta}\end{aligned} \tag{8.112}$$

Define

$$\bar{c}_0 = \min\{\bar{c}_1, \bar{c}_2, \ldots, \bar{c}_\rho\} \tag{8.113}$$

$$\lambda_0 = \min\{l_\varepsilon / \lambda_{\max}(P), \bar{c}_0, 1\} \tag{8.114}$$

Then, from (8.113) and (8.114), (8.112) becomes

$$\dot{V}(t) \leq -\lambda_0 V(t) + \upsilon_0 \tag{8.115}$$

Further, one has

$$0 \leq V(t) \leq \frac{\mu_0}{\lambda_0} + [V(0) - \frac{\mu_0}{\lambda_0}]e^{-\lambda_0 t} \leq \frac{\mu_0}{\lambda_0} + V(0) \tag{8.116}$$

Let $\alpha_0 = \upsilon_0/\lambda_0 + V(0)$. Then, from (8.106) and (8.116),it follows

$$||\varepsilon|| \leq \sqrt{\alpha_0/\lambda_{\min}(P_i)},\ ||z|| \leq \sqrt{2\alpha_0},\ ||\tilde{\theta}|| \leq \sqrt{2\eta_\theta\alpha_0},\ ||\tilde{\gamma}|| \leq \sqrt{2\eta_\gamma\alpha_0},$$
$$||\tilde{b}_y|| \leq \sqrt{2\eta_b\alpha_0},$$

which means that the closed-loop system is asymptotically stable and all the closed-loop signals converge asymptotically to a small neighborhood of the origin.

Because $z_q = [z_{1,q}, \ldots, z_{N,q}]^T$, it follows from Lemma 8.1 that $||z_q|| \leq \sqrt{2\alpha_0}$. Then, one has, for $q = 1, \ldots, n$,

$$\epsilon_q \leq \sqrt{2\alpha_0}/\underline{\varsigma}(B+L),$$

which implies that the synchronization between the leader and followers has been achieved with the neighborhood and tracking errors are asymptotically converged to the small neighborhood of the origin.

The above analysis is summarized by the following theorem.

Theorem 8.1 *Consider the leader-following system (8.1) and (8.2) with Assumptions 1–4. If the distributed control and adaptive laws (8.98)–(8.101), by appropriately choosing the parameter* $h, r, \eta_\theta, \eta_\gamma, \eta_b$, *then the synchronization between the leader and followers is achieved with the neighborhood and tracking errors are asymptotically converge to the small neighborhood of the origin* Ω *defined as*

$$\Omega = \left\{ \begin{array}{l} (\varepsilon, z, \tilde{\theta}, \tilde{\gamma}, \tilde{b}_y, \chi_q) | ||\varepsilon|| \leq \sqrt{\alpha_0/\lambda_{\min}(P)}, \\ ||z|| \leq \sqrt{2\alpha_0},\ ||\tilde{\theta}|| \leq \sqrt{2\eta_\theta\alpha_0}, \\ ||\tilde{\gamma}|| \leq \sqrt{2\eta_\gamma\alpha_0},\ ||\tilde{b}_y|| \leq \sqrt{2\eta_b\alpha_0}, \\ \epsilon_q \leq \sqrt{2\alpha_0}/\underline{\varsigma}(B+L), q = 1, \ldots, n \end{array} \right\} \tag{8.117}$$

Proof The conclusions can be obtained easily from the above analysis. The detailed proof is omitted here.

8.4 Simulation

In this section, let us consider a power system with 4 buses. In the system, each bus denotes a follower. In addition, a leader is also considered. Among the buses and leaders, their communication is represented by a graph presented in Fig. 8.3. For buse i ($i = 1, \ldots, N$), it has the following form

$$m_i\ddot{\delta}_i(t) + d_i\dot{\delta}_i(t) - P_{mi}(t) = -\sum_{j\in N_i} P_{ij}(t)$$

where d_i, m_i and δ_i respectively the damping coefficients of generators, inertia coefficients of motors and phase angle; P_{ij} represents the active power flow from v_i to v_j, P_{mi} represents the mechanical input power, and N_i represents the neighborhood set of node i. Notice that, buses j and i can share a common route.

Define $\xi_i(t) = \delta_i(t)$ and $\zeta_i(t) = \dot{\delta}_i(t)$, the bus i becomes

$$\begin{aligned}\dot{\xi}_i(t) =&\zeta_i(t)\\ \dot{\zeta}_i(t) =& -\frac{d_i}{m_i}\zeta_i(t) - \frac{1}{m_i}\sum_{j\in N_i}[\omega_{ij}^1 \cos(\xi_i(t) - \xi_j(t)) +\\ &\omega_{ij}^2 \sin(\xi_i(t) - \xi_j(t))] + \frac{P_{mi}(t)}{m_i}\end{aligned}$$

Further, define $v_i = [v_{i,1}, v_{i,2}]^T = [\xi_i, \zeta_i]^T$, the bus can described in the form (8.1),

$$\begin{cases}\dot{v}_{i,1} = v_{i,2} + f_{i,1}(y_i) + d_{i,1}\\ \dot{v}_{i,2} = u_i + f_{i,2}(y_i) + d_{i,2}\\ y_i = v_{i,1},\ i = 1, \ldots, 4\end{cases} \tag{8.118}$$

where $f_{i,1}(y_i) = 0.5v_{i,1}\sin(v_{i,1})$, $f_{1,2}(y_1) = -3x_{1,1}^2 + 2v_{1,1}\cos v_{1,1}$, $f_{2,2}(y_2) = 0.5v_{2,1} + v_{2,1}\sin v_{2,1}$, $f_{3,2}(y_3) = -v_{3,1}^2 - 0.1\sin^2 v_{3,1}$, $f_{4,2}(y_4) = 2v_{4,1}\cos v_{4,1}$, $d_{i,1} = 0.01\sin t$ and $d_{i,2} = 0.01\cos t$. In this simulation, the sensor faults are set as $y_{i,f} = y_i + 2$, $i = 1, \ldots, 4$. In addition, for the leader, $f_0(t) = 0.5\sin(t)$ in this work. From Fig. 1, we know that

$$G = \begin{bmatrix} 0\,1\,0\,0\\ 0\,0\,0\,0\\ 0\,1\,0\,0\\ 0\,0\,1\,0\end{bmatrix}, L = \begin{bmatrix} 1 & -1 & 0 & 0\\ 0 & 0 & 0 & 0\\ 0 & -1 & 1 & 0\\ 0 & 0 & -1 & 1\end{bmatrix}, B = \begin{bmatrix} 0 & & & \\ & 1 & & \\ & & 0 & \\ & & & 0\end{bmatrix}$$

Initial conditions are set as: $v_1(0) = [0.5, 0]^T$, $v_2(0) = [-0.3, 0]^T$, $v_3(0) = [0.15, 0]^T$, $v_4(0) = [-0.1, 0]^T$, $\xi_i(0) = [0, 0]^T$, $\lambda_i(0) = [0, 0]^T$, $c_{i,1} = 13.c_{i,2} = 5$, $k_{i,1} = 10$, $k_{i,2} = 30$.

The simulation results are shown in Figs. 8.4, 8.5 and 8.6. From Fig. 8.4, it can be found that good tracking performance has been obtained. Meanwhile, it can be seen from Fig. 8.5 that each tracking error converges asymptotically to a small neighborhood of the origin. For the followers' control inputs, their boundedness is shown in Fig. 8.6.

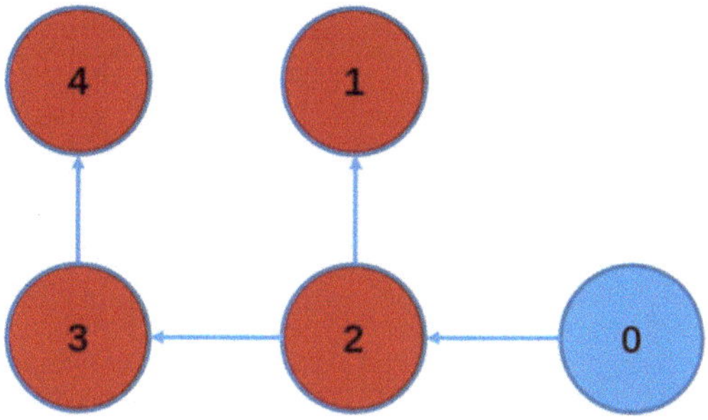

Fig. 8.3 Topology of the graph

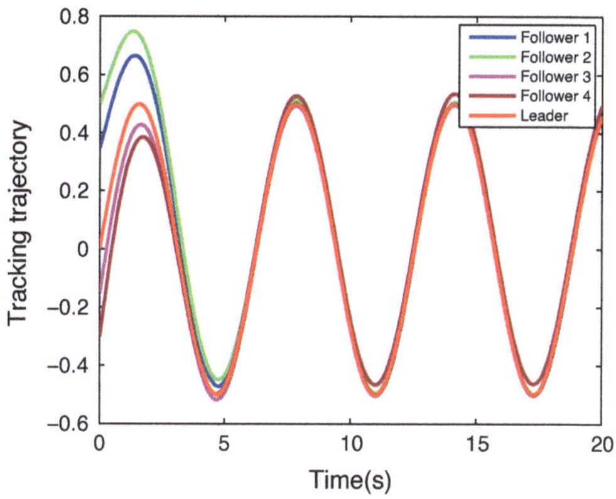

Fig. 8.4 Follower outputs y_i and desired signal y_0

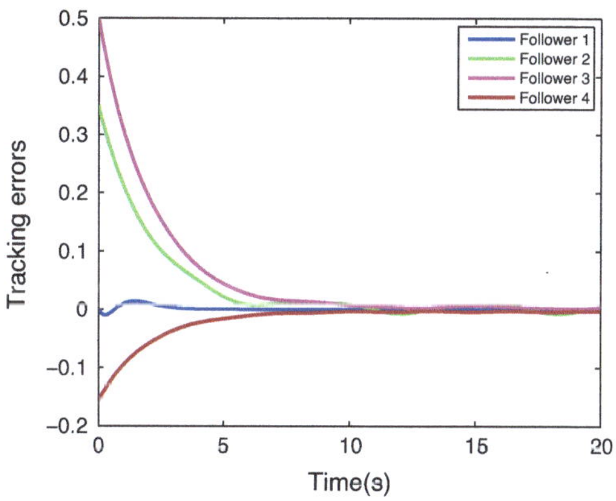

Fig. 8.5 Follower tracking errors

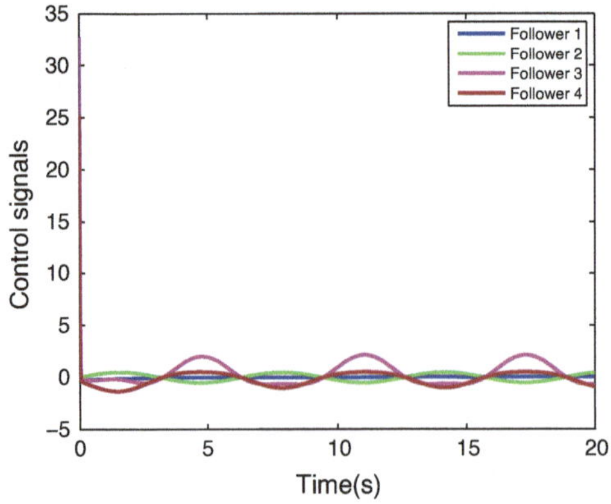

Fig. 8.6 Follower control inputs

8.5 Conclusion

In this chapter, a distributed output-feedback control problem is considered for a class of nonlinear uncertain leader-following systems. Considering the sensor faults, combining backstepping method and other adaptive control techniques, a distributed FTC strategy scheme has been designed such that the synchronization between the leader and followers has been achieved with the neighborhood, and tracking errors are asymptotically converge to the small neighborhood of the origin. It has been proved that the proposed control scheme can guarantee that all signals in the closed-loop systems are uniformly ultimately bounded. Moreover, the tracking errors can converge to a small adjustable neighborhood of the origin. The simulation results demonstrate the effectiveness of the control scheme in this chapter.

References

Cheng L, Hou Z, Tan M, Lin Y, Zhang W (2010) Neural-network-based adaptive leader-following control for multiagent systems with uncertainties. IEEE Trans Neural Netw 21(8):1351–1358

Das A, Lewis F (2010) Distributed adaptive control for synchronization of unknown nonlinear networked systems. Automatica 46(12):2014–2021

Defoort M, Floquet T, Kokosy A, Perruquetti W (2008) Sliding-mode formation control for cooperative autonomous mobile robots. IEEE Trans Ind Electron 55(11):3944–3953

Fax JA, Murray RM (2003) Information flow and cooperative control of vehicle formations. IEEE Trans Autom Control 49(9):1465–1476

Hang C, Zhang T (2002) Stable adaptive neural network control. Kluwer Academic, Boston

Krstic M, Kanellakopoulos I, Kokotocic PV (1995) Nonlinear and adaptive control design. Wiley, New York

Li Y, Yang G, Tong S (2019) Fuzzy adaptive distributed event-triggered consensus control of uncertain nonlinear multiagent systems. IEEE Trans Syst Man Cybernet: Syst 49(9):1777–1786

Pan W, Yua Y, Sandber H, Gonçalves J, Stan G (2015) Online fault diagnosis for nonlinear power systems. Automatica 55:27–36

Peng Z, Wang D, Zhang H, Sun G (2014) Distributed neural network control for adaptive synchronization of uncertain dynamical multiagent systems. IEEE Trans Neural Netw Learn Syst 25(8):1508–1519

Shen QK, Shi P (2015) Distributed command filtered backstepping consensus tracking control of nonlinear multiple-agent systems in strict-feedback form. Automatica 53:120–124

Shen QK, Jiang B, Cocquempot V (2012) Fault-tolerant control for T-S fuzzy systems with application to near-space hypersonic vehicle with actuator faults. IEEE Trans Fuzzy Syst 20(4):652–665

Shen QK, Jiang B, Shi P, Zhao J (2013) Cooperative adaptive fuzzy tracking control for networked unknown nonlinear multiagent systems with time-varying actuator faults. IEEE Trans Fuzzy Syst 22(3):494–504

Shen QK, Jiang B, Shi P (2014) Adaptive fault diagnosis for T-S fuzzy systems with sensor faults and system performance analysis. IEEE Trans Fuzzy Syst 22(2):274–285

Wang W, Tong S (2018) Adaptive fuzzy bounded control for consensus of multiple strict-feedback nonlinear systems. IEEE Trans Cybernet 48(2):522–531

Wang X, Yadav V, Balakrishnan S (2007) Cooperative UAV formation flying with obstacle/collision avoidance. IEEE Trans Control Syst Technol 15(4):672–679

Xia X, Zhang T, Yi Y, Shen QK (2016) Adaptive prescribed performance control of output feedback systems including input unmodeled dynamics. Neurocomputing 190:226–236

Xiao B, Hu Q, Zhang Y (2012) Adaptive sliding mode fault tolerant attitude tracking control for flexible spacecraft under actuator saturation. IEEE Trans Control Syst Technol 20(6):1605–1612

Yoo SJ, Kim TH (2015) Distributed formation tracking of networked mobile robots under unknown slippage effects. Automatica 54:100–106

Zhai D, An L, Li J, Zhang Q (2016) Adaptive fuzzy fault-tolerant control with guaranteed tracking performance for nonlinear strict-feedback systems. Fuzzy Sets Syst 302:82–100

Zhang H, Lewis F (2012) Adaptive cooperative tracking control of higher-order nonlinear systems with unknown dynamics. Automatica 48(7):1432–1439

Zhang Y, Li H, Sun J, He W (2019) Cooperative adaptive event-triggered control for multiagent systems with actuator failures. IEEE Trans Syst Man Cybernet: Syst 49(9):1759–1768

Chapter 9
Conclusion and Future Research Directions

9.1 Conclusions

With the wide application, multi-agent systems including leader-following systems are getting bigger and bigger, and more and more single agents are involved. On the other hand, with the development of science and technology, modern control systems (single-agent systems), such as flight control systems, also become more and more complex, which means an increasing number of actuators and sensors. In practical applications, these physical components may become faulty. The faults that occurred in the components may cause system performance deterioration and even instability, which can further may result in catastrophic accidents. Communication networks and their links also become faulty Due to the harsh environment, enemy attacks, etc. What's more worst, the faults that occurred in an agent can affect the other agents in the whole multi-agent system. Hence, the study of fault-tolerant control for multi-agent systems has important theoretical value and practical application significance. This book focuses on the issues of adaptive fault-tolerant control for multi-agent systems including leader-following systems. The main contributions presented in this book include

1. Different from most of the existing works, where the input nonlinearities considered in this book were not considered, uncertain nonlinear MASs with input nonlinearities are widely studied, and a distributed adaptive NN-based control scheme is proposed to guarantee that output consensus is achieved, which means that the results obtained in this book have more practical potential.
2. Novel distributed adaptive control schemes for all the followers do not require the assumption that the upper bounds of optimal approximation errors and external disturbance should be known exactly.
3. A decision threshold for fault detection is defined and applied on an online computable fault indicator and not on an asymptotic value of a criterion which is not available in practice. The decision algorithm is thus more practical than in other existing works in literature.

Q. Shen, *Distributed Fault-Tolerant Consensus Control of Leader-Following Systems*, Intelligent Control and Learning Systems 11,
https://doi.org/10.1007/978-981-99-7426-9_9

4. A new FTC scheme proposed in the book not only can minimize time delay induced by fault diagnosis but also can reduce the adverse effect on system performance. In addition, it does not require an additional fault isolation model, which is necessary in the traditional active FTC scheme.
5. Supervisor control strategy is first used to control MASs, which ensures the boundedness condition is satisfied and makes the stability analysis of the closed-loop systems more complete and rigorous.
6. Two different time-varying communication faults (transmission nonlinearity) models are introduced, which can deal with a wider class of the transmission nonlinearities in practical applications. two novel distributed adaptive consensus control schemes are, respectively, proposed for a class of leader-following systems with the above transmission nonlinearities, where two compensation algorithms for the transmission nonlinearities are, respectively, given.
7. In the presence of the condition that the input signals of radial basis function neural networks deviate from the true values due to sensor faults, the analytical expression of output layer of RBF neural networks is obtained by analyzing the structure of the neural networks.
8. The proposed control scheme combines the K-filters design method that only uses the available neighborhood output estimation error information and neural network backstepping control techniques, and the cooperative fault-tolerant tracking control problem is solved.

9.2 Future Research Directions

Fault-tolerant control for multi-agent systems is a very hot research topic that has important both academic meaning as well as practical one. This book has presented several recent results and applications on this topic. For the fault-tolerant control for multi-agent systems, although it has been extensively studied, there also exist still a great number of crucial and fundamental issues to be exploited, which are open and challenging. To the best of our knowledge, some problems are listed as follows:

1. **Isolation of the faulty agents**
   In MASs, there exist many agents, and an agent can affect the other agents via communication network. Hence, before isolating the local faults that occurred in agents, the designers must know whether an agent is faulty or not. Based on the condition that the faulty agents are known to the designers, the corresponding fault isolation algorithms can be constructed. Hence, it is necessary and important to distinguish the faulty agents from the health agents, which is called the isolation of the faulty agents. However, although the fault-tolerant control of MASs is widely investigated, most of the existing results in literature do not focus on the isolation of the faulty agents. Obviously, the isolation of the faulty agents still is an open issue.

2. **Isolation of the local faults**
   As stated in the previous chapters, the faults occurred in MASs can be divided into two categories: local and global faults. Like the faults in single-agent systems, the local faults of MASs also occur inside agents. In some senses, the local fault isolation procedures of MASs are similar to the fault isolation of single-agent systems. However, comparing the isolation of single-agent systems, the local fault isolation of MASs is more complex. Furthermore, in general, the fault isolation methods for single-agent systems cannot be directly applied in the fault isolation of MASs, and novel ideas should be aroused to design the fault isolation methods. Hence, the local fault isolation of MASs is worth to be deeply studied.
3. **Isolation of the global faults (the faulty communication links)**
   In MASs, there are many links in the communication network. Furthermore, the weights impacted on the links are different, which implies that different agents have different effects on the whole MASs. In fact, how to distinguish the faulty links from the health links is very important. However, up to now, it is not studied enough. Hence, the communication fault isolation of MASs is necessary and worthy of investigation.
4. **Modeling of communication fault**
   In this book, the faults that occurred in communication links are simplified and regarded as special sensor faults. Thus, the faults have similar fault categories: gain and bias faults. The purpose of the simplification is to simplify the difficulties of fault compensation algorithms, and the fault modeling is not enough considered the physical properties of communication links. Hence, for the communication link faults, how to understand the physical properties and propose more suitable fault models still is open and challenging.

MIX
Papier aus verantwortungsvollen Quellen
Paper from responsible sources
FSC® C105338
FSC
www.fsc.org

If you have any concerns about our products,
you can contact us on
ProductSafety@springernature.com

In case Publisher is established outside the EU,
the EU authorized representative is:
Springer Nature Customer Service Center GmbH
Europaplatz 3, 69115 Heidelberg, Germany

Printed by Libri Plureos GmbH
in Hamburg, Germany